WHAT THEY ARE SAYING ABOUT VIDEO COMPRESSION FOR MULTIMEDIA

Though Jan's book explains the inner workings of digital video, it is not just for the experts. The information is essential for *anyone* doing video capture. By using the techniques he recommends, you can be sure that your video will be done right the first time.

> —**Barry Hudson, Multimedia Planner**
> **Westinghouse**

This is the best, most comprehensive treatment of digital video compression techniques for creators of interactive multimedia products and services. Highly recommended.

> —**Doug Millison, Editor-in-Chief**
> *Morph's Outpost on the Digital Frontier*

You might think that this is more than any developer would want to know about digital video compression. Not so. Understanding the technology is essential when planning, purchasing and problem solving, and Jan Ozer's discussion is both thorough and authoritative. It is also an invaluable reference for on-the-fly questions that the manuals never answer. We've distilled tip sheets for everything from shooting to compression settings. I only hope Mr. Ozer will keep us supplied with updates on this rapidly changing technology.

> —**Jan Diamondstone, President**
> **Interactive Design**

Jan Ozer's landmark book comes at an important moment now in the pre-Kitty Hawk days of digital multimedia. For the practitioner, this book is an especially valuable guide, a reliable "how-to" without tears and fears. For those beginning to plan for multimedia applications, a finer introduction to digital video compression, the ultimate enabling technology, is difficult to imagine.

> —**William J. Caffery, Vice President**
> **Advanced Technology Gartner Group, Inc.**

Video Compression
for Multimedia

Video Compression

for Multimedia

Jan Ozer

CD-ROM
ENCLOSED

AP PROFESSIONAL

Boston   San Diego   New York
London   Sydney   Tokyo   Toronto

AP PROFESSIONAL
955 Massachusetts Avenue, Cambridge, MA 02139

An Imprint of ACADEMIC PRESS, INC.
A Division of HARCOURT BRACE & COMPANY

United Kingdom Edition published by
ACADEMIC PRESS LIMITED
24–28 Oval Road, London NW1 7DX

Library of Congress Cataloging-in-Publication Data

Ozer, Jan, 1955-
     Video compression for multimedia / Jan Ozer.
          p.   cm.
     Includes index.
     ISBN 0-12-531940-1 (acid-free paper)
     1. Multimedia systems. 2. Video compression. I. Title.
QA76.575.O94   1994
006.6--dc20 94-32561
                                                    CIP

Printed in the United States of America
94 95 96 97 98 IP 9 8 7 6 5 4 3 2 1

To Barbara: Wife and best friend

To Margo and Jack Ozer: Parents and other best friends

CONTENTS

INTRODUCTION

THE LAY OF THE LAND

Thanks for buying this book. This chapter will present an overview of the digital video market and describe what we will and won't be covering in the book.

Software vs. Hardware

Digital video technologies divide into two major categories, those that require specialized hardware for decompression, and software-only technologies that decompress using the host CPU. Examples of software-only video are Video for Windows, Microsoft's video architecture and applications suite, QuickTime by Apple and other technologies that play back video without hardware assistance. Examples of hardware-only systems include a technology called MPEG, a standard promulgated by the Motion Pictures Expert Group, and proprietary technologies such as DVI and Motion JPEG (Fig I.1).

Hardware-assisted technologies typically offer up to full screen video at 30 frames per second, with video quality approaching "broadcast," or television quality. The obvious price for this quality is the hardware necessary to play back the video, usually a separate printed circuit card installed in every playback station. The compression side of hardware-

Figure I.1 Decision 1 for multimedia developers—hardware-assisted vs. software-only playback

Digital Video Market Segments	
	Examples
Software only	Video for Windows QuickTime for Windows Fractal Video
Hardware-Assisted	MPEG Digital Video Interactive (DVI) Motion JPEG

assisted technologies is usually more expensive and less accessible—often in the $5,000–20,000 range.

Applications for hardware-assisted video tend to be either very large or very small. MPEG is achieving some success in the home games market and seems destined to be the technology of choice for "set-top boxes" carrying cable movies into the home. At the other end of the spectrum, Motion JPEG products are frequently used in small, closed applications such as small training departments and kiosks.

Software-only technologies offer smaller resolutions, usually 320x240, and typically display about 15 frames per second, or about half the rate of television. Video quality ranges from downright awful to just under broadcast quality for some clips in some applications. However, software-only playback is essentially free—your video consumers don't have to purchase or install special hardware. For this reason, software-only video is used across a broad spectrum of applications.

According to some reports, the Seventh Guest, the current "hot" multimedia title, has sold over 4,000,000 copies. Mid-range applications for software-only video include corporate training to network video-on-demand systems, and applications get as small as capturing family events for digital posterity.

Which is right for you? It depends entirely upon the application (Fig. I.2). If your application won't be widely distributed, or is so special that you can require your target customers to purchase and install special hardware, hardware-assisted play-back may be appropriate. On the other hand, recent improve-ments in software-only technology make it worthy of consider-ation for almost all applications.

Hardware-assisted technologies were formulated when 80386-powered, ISA bus systems prevailed. As processor, bus and video graphics technologies advance, the overall pro-cessing power in a typical system exceed the combined power of a 386 and video decompression board. As the line between hardware-assisted video and software-only video blurs, software-only video becomes more attractive, because it performs as well as hardware-assisted without the risk of obsolescence.

Our Focus

This book will focus primarily on the software-only technolo-gies. We will briefly discuss MPEG in Chapter 15, and touch on Motion JPEG products in Chapter 8 when discussing cap-ture cards. Here's how the rest of the book lays out.

Figure I.2 Characteristics and applications of hardware-assisted and software-only video

Evaluating your Alternatives

Option	Characteristics	Applications
Software-only	Acceptable video quality "Free" decompression No obsolescensce risk	Most consumer multimedia titles Large Scale training Virtually limitless
Hardware Assisted	Limited target market Near broadcast quality $$$$ and installation hassles Obsolescence risk	Home "set top" boxes Small, "closed" applications like kiosks Some multimedia titles

Theory and Overview

CHAPTER 1—INTRODUCTION TO DIGITAL VIDEO
What's the big deal about digital video? Everything! It's all huge. Here we'll overview the capture and playback process, and you'll see why digital video eats kilobytes for breakfast, megabytes for lunch and gigabytes for dinner. Better order your new hard drive now.

CHAPTER 2—VIDEO COMPRESSION
This chapter describes how compression works. Not just for the propeller heads and compression junkies (like me), but critical for everyone working with digital video. You'll learn how to film for compression and how compression options we'll use later affect video quality and display rate.

CHAPTER 3—INTRODUCTION TO VIDEO FOR WINDOWS
We'll spend most of our time working in Video for Windows. This chapter explains what it is, how it works and where it's going.

Operation

CHAPTER 4—SOUND SYNCHRONIZATION
Nothing's more distracting than poorly synchronized video. While not strictly a compression function, learning how synchronization works is essential to producing high-quality video.

CHAPTER 5—PLAYBACK PLATFORM CONSIDERATIONS
Software-only video plays to the level of the host hardware, and you shouldn't capture or compress without knowing how the configuration of your target system will impact video display.

Here we'll see how processors, bus technologies, CD-ROMs and advances in video technologies affect video playback.

CHAPTER 6—TOUR DE CODEC

All video technologies have their uses. In this chapter we'll see how, where and why to use technologies like Indeo and Cinepak—and where not to.

Your Capture Dollar

CHAPTER 7—ANALOG OVERVIEW

You've got a budget and you want to spend it where it brings the most quality to the final compressed video. Analog video comes in many flavors, each with a different price tag. Does investing in quality at the front end create better video? You'll see for yourself in this chapter.

CHAPTER 8—CAPTURE KARMA II

What features should you look for in a video capture card? Which current card gives you the most bang for your capture buck? Find out here.

CHAPTER 9—SURVIVING YOUR CAPTURE BOARD INSTALLATION

> "Multimedia Publishing, here I come!!!!"
> Jan Ozer, 6/93

> "Life's a bitch . . . then you try to capture video"
> Jan Ozer, 6/94

There's a term for trying to install a capture board without reading this chapter. It's called Plug and Pray. The time you spend on this chapter will pay immediate dividends when you install your new capture card.

Capture, Preprocessing and Compression

CHAPTER 10—VIDEO CAPTURE

Video capture is the first step in the digital process. Here's where you select frame rate, resolution and other video characteristics. This chapter presents statistics and sample videos to aid your selection and describes how to maximize capture quality. Step by step, we also describe the software controls and hardware links needed to capture in both step-frame mode and real time.

CHAPTER 11—PREPROCESSING FOR COMPRESSION

Now that your video is on disk, it's time to get it ready for compression. We'll describe how to trim the video to its final form, filter out noise, and how advanced functions such as transitions affect how well the video will compress.

CHAPTER 12—VIDEO FOR WINDOWS COMPRESSION CONTROLS

The chapter we've been waiting for—finally ready to compress! Here we apply compression theory to select the right compression technology and parameters and produce the highest-quality video file.

CHAPTER 13—VIDEO FOR WINDOWS DEVELOPMENT TIPS

"Working with Video for Windows is like living in a haunted house. You can't predict when strange things will happen, but the longer you live there, the more likely some seemingly innocent step will lead to totally bizarre results."

Using the wrong palette can drop your display rate by 66%. Ditto for window type and size. Read this chapter and find out where the skeletons are in Video for Windows.

CHAPTER 14—YOUR CAPTURE STATION

This chapter describes how to configure your primary capture and compression station.

CHAPTER 15—MPEG

You've heard all the MPEG hoopla. Here we'll take a (mostly) objective view of the technology and products that implement it. We'll also predict where MPEG is going, and where it isn't.

Summary

Overall, after the initial theory, the book organization tracks the video creation process from capture to compression (Chapters 7-12). Chapters 9-12 are more workshops than theory, and are best read while doing the work.

All chapters are fairly modular, so you can jump in anywhere without losing synch. Each chapter starts with a brief description of the contents and ends with a summary, so it should be pretty easy to tell whether the information is immediately relevant or not.

What We Don't Cover

Mac. Mac envy. How wonderful we all are because we're working with video. Capture boards that cost under $400. CD-ROM mastering. How to make your movies look like the *Terminator* or *Jurassic Park*. Morphs. Programming with C, C++, Assymetrix ToolBook or Visual Basic.

If you're a Mac person, or need to extensively explore cross-platform issues, you might try Nels Johnson's book, *How to Digitize Video*, published by John Wiley and Sons. For a broad overview of digital video with a focus on programming, I recommend you pick up *PC Video Madness*, by Ron Wodaski from SAMs publishing. Both books are widely available.

I'll end the beginning with a story. I was on a press tour, visiting a West Coast magazine, meeting with a technical analyst and a reporter covering the business side of the announcement. The reporter hadn't yet arrived. The president of my company had just published his second book on an obscure branch of geometry, and the analyst was thrilled that I had brought her a copy.

When the reporter came in, the analyst raved about the book and asked the reporter if she, too, wanted a copy. The reporter looked at the title and said, "No—why don't you just read it and tell me how it ends."

I hope you'll want to read this one and see for yourself.

How to Read This Book

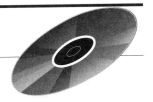

The CD-ROM

The CD-ROM should be considered an integral part of the book. The CD-ROM is arranged in subdirectories by chapter name. Most of the videos referred to in the book are included in their respective chapter by figure number. For example, Fig12_12.avi would be the source of Figure 12.12 in the book. The printed page cannot completely and accurately represent video, so we urge you to follow along. The CD symbol in the left margin is an additional indicator of which figures can be found on the CD-ROM.

Install the CD-ROM by running Setup.exe from the CD-ROM root directory. All required drivers are contained on the disk and will be installed automatically, along with VCS Play, an application to help you view and compare the files.

VCS Play

Included with the CD-ROM is a truncated version of the VCS Play application that normally retails for $100. The version on the CD-ROM operates just like the retail version except that printing is not available and the CD-ROM itself must be in your drive to load the program. You can remove the CD-ROM from your disk after VCS Play is loaded and it will work with other Video for Windows files.

Also included is a viewer from Xing Technology for Scalable MPEG files. These files will not play on VCS Play, so you should use Xing's viewer to play back the MPEG files in Chapter 6.

The VCS retail version also contains a complete catalog of videos illustrating the various codecs using four benchmark sequences ranging from Talking Head to High Action clips. We update the codecs periodically for new codecs and new releases of current codecs.

A Quick Tour of VCS

This tour will introduce you to basic VCS features.

1. *Load VCS*—After seamless installation (gulp) of VCS Play, XING's MPEG Viewer and Video for Windows, double click on the VCS icon. You'll start in the File Selection screen.

2. *Directory Structure*—select your CD-ROM drive. You should see the directory structure showing the individual chapters separated into different subdirectories.

3. *Select File*—double click on chap_6 subdirectory and touch Fig6_2.avi. Note the file information that appears below. This information is extremely helpful in locating files after you've started compressing your own.

4. *Getting to know VCS*—Double click on the file name to load the file. While loading VCS Play scans the first 100 frames checking for key frame information—this takes a few seconds for files on CD-ROM.

   a. Move mouse over control bar just under video. Note instructions in lower left hand corner (very trendy!!).

   b. Place mouse cursor on number 18331 and click left button. Keep clicking until bored. Then use right mouse key to return to normal resolution. You'll find zooming invaluable when comparing the various videos included on the disk.

c. Press the "ear" in the lower left hand corner. Note audio statistics. Press F1 for help screen and select audio statistics to view help files explaining audio stats. Exit help and click "ear" again to return to main screen.

d. Press the "eye" on the bottom toolbar. Press "compute." This calculates the "frame profile." This takes a few moments when running on a CD-ROM. Press F1, search for "Frame Profile." Go to "Details on Frame Profile" to read up on its significance.

e. Click the "view" button. This should toggle you back to the video screen. Use the double right arrow key to step forward through the video. The wavy blue/green line is the per second data rate, very useful when trying to determine if a video will play from a CD-ROM drive. More on this in Chapter 7.

5. *Play the video*—press the "play" key located next to the speaker button on the upper toolbar.

To load a file in the left hand window, press the yellow folder in the bottom left hand corner of the right hand side of the screen. The folder on the left loads another video file on that side.

That's about it. We hope you enjoy VCS Play and the book.

ACKNOWLEDGMENTS

I would like to thank the following companies and people for their contributions to this book.

Equipment and Access: IBM Skill Dynamics, for three days unfettered access to their analog studios; Media Management, for more or less continuous access to their facilities; Intel, for providing frequent compression services, equipment, beta code, advice and information; Horizons Technology, for capture and compression services above and beyond the call; Uriah Heap, for capture services and testing assistance, Interactive Design, capture services and technical assistance.

Editorial Assistance: Heidi Carson, Jan Diamondstone, Dave Duberman, Jackie Gavron, Dave Guenette, Barry Hudson, Cliff Johnson, Matt Lake, Jerry McFaul, Reed McMillan, Rockley Miller, Doug Millison, Steve Pitt, Dave Trowbridge

Industry Information and Gossip: Bill Caffery, Bill Coggshall, Chris Eddy, Steve Edelson (aka Steve, the Video Genius), Jennifer Edstrom, Steve Griffin, Don Harris, Jon Hill, Peter Jacso, Jim Karney, Jeff Mace, Kevin O'Connell, LJ Purtee, Rick Purtee, Robin Raskin, Erica Schroeder, (hears all, sees all, knows all), Ron Wadowski.

INTRODUCTION TO DIGITAL VIDEO

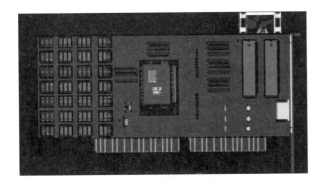

INTRODUCTION TO DIGITAL VIDEO

2

IN THIS CHAPTER

The term digital video refers to video playing on a computer in digital format. It does not include analog video playing through a special add-in card, or animation. This section covers two topics. First, we'll briefly describe the capture process, where video and sound are converted from analog to digital format. This topic is covered in detail in Chapter 10. Then we'll discuss the storage and bandwidth problems this conversion creates.

Concepts covered in this chapter are used throughout the book, and a working knowledge is essential to your understanding of future chapters. More importantly, we'll define many of the acronyms, slang and technojargon used in and around video and multimedia. So if you want to look hip and stay synched to the digital flow, click "OK" and read on.

Welcome to the 'hood.

WHAT IS VIDEO

Video is a stream of data composed of discrete frames, usually including both audio and pictures. Television signals in the United States have 30 discrete frames per second, while European stations broadcast 25 frames per second. Most movies are filmed and played back at 24 frames per second.

When originally captured on film, whether by your father's old 16 mm camera or by the newest Hi-8 camera, both the pic-

Figure 1.1 Analog sine wave

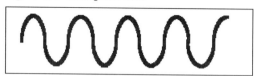

tures and sound are *analog* in format. For me, analog was always a pretty confusing term. Definitions like "the representation of information by variable physical quantities such as the size of electrical voltages—voltages which are analogous to the original data" just didn't compute.

The term finally started to make sense when an engineer took his finger and drew a curvy sine wave like that shown in Figure 1.1. Analog is a naturally continuous signal, with breadth and depth. There are no pixels or other fixed boundaries, no precise values. The signal is virtually infinitely magnifiable, which lets you take a 1-inch slide and blow it up to billboard size without loss of quality or pixelation.

In contrast, computers are digital devices that process all information as zeros and ones. A picture on your computer isn't continuous. It's a collection of pixels, each with a precise value, usually defined as some blend of red, green and blue. If you blow a computer picture up to billboard size you still have the same number of pixels—they're just bigger. So your billboard ends up looking like those huge screens at stadiums that look great from far away but resemble color connect-o-dot drawings from close up.

When you convert video from analog to digital format, or *digitize* the video, you take in the analog signal, divide it into discrete pixels, and assign a precise value to each pixel. In the old days of 2- and 4-bit graphics cards and monitors that displayed only 16 different colors, pictures looked awful. There just weren't enough colors to simulate the analog event.

Today's 24-bit video cards offer over 16 million colors, more than enough to fool the eye into believing that the digital video shown on the screen is actually a continuous analog event. But we're getting ahead of ourselves. Let's look at the hardware.

VIDEO CAPTURE—OVERVIEW

OK. You've got some analog footage to digitize. Maybe some clips from your latest annual meeting, customer testimonials or maybe just some home movies. The first thing you need is an *analog source* to input the analog video signal into the computer. Generally, any device that can feed a signal to a television can be your analog source.

Then you need a *capture card* or *frame grabber,* which is a printed circuit card installed in your computer. As a point of reference, capture cards that operate at 30 frames per second are called "real-time" capture cards. Most capture cards come with software that controls the capture process. For example, when working on the Windows platform, the software is usually Microsoft's Video for Windows.

Most analog devices feed 30 frames per second into the computer at resolution of 640x480 pixels. Most video destined for compression is captured at 15 frames per second at a resolution of 320x240. So the capture card's first task is shrinking the video to the target resolution, and reducing the frame rate by dropping every other frame. These processes are both forms of *scaling*, where the characteristics of the analog signal are scaled back for use on the digital platform.

After scaling, the card digitizes the video by dividing the frame into pixels and assigning a digital value to each pixel. This process is shown in Figure 1.2.

Video—You can view a live demonstration of this process by double-clicking on CAPTURE.AVI in the chap_1 subdirectory.

Most capture cards don't digitize audio, so you'll also need a sound card, another printed circuit card installed in your computer. Like the capture card, the sound card takes in the analog audio signal and assigns a digital value to specified points in the audio stream (see Figure 1.3).

When video and audio are captured on separate devices, the digitization software synchronizes the two digital files. For example, VidCap, Video for Windows' capture utility, automatically synchronizes the video and audio data and creates a combined file by interleaving audio and video data together. In fact, Video for Windows' AVI file format stands for audio/video interleaved.

Figure 1.2 Capture board digitizing frames from a laserdisc

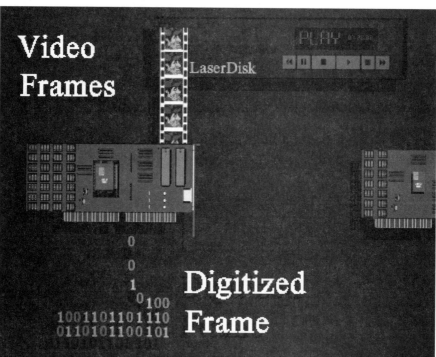

The interleaved file format not only helps synchronize the file, it's also useful for playback from CD-ROMs. That's because it's more efficient to retrieve one interleaved data stream than to perform the multiple seeks required to retrieve separate audio and video streams.

BANDWIDTH DEFINED

The term **bandwidth** is used in two ways. First, it is a measure of a device's capacity, or ability to move information to or from a device, system or subsystem. Usually this is measured in terms of quantities of data per second. For example, a single spin CD-ROM player can transfer information from the CD-

ROM into the computer at a maximum rate of 150 kilobytes per second (kB/s). Accordingly, the bandwidth of a single spin CD-ROM is 150 kB/s.

Bandwidth is also used to describe the data flow necessary to support a process such as video playback. A 30-second video file that is 4.5 megabytes (MB) in size has a bandwidth of 150 kB/s, because data must be transferred at that rate to play the file without interruption. This usage is a measure of requirement. When used in this manner, bandwidth is used interchangeably with the term *data rate*.

As you would expect, problems occur when a video's data rate exceeds the bandwidth capacities of any system component. Since the data rate of raw digital video exceeds the transfer capacities of almost all computer system components,

Figure 1.3 Sound board digitizing audio from a laserdisc

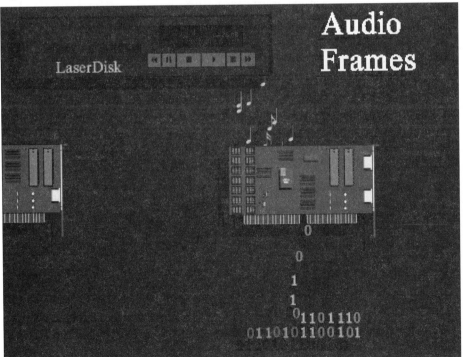

bandwidth is an issue that must be addressed early and often in the video capture and creation process.

VIDEO BANDWIDTH

Let's analyze the bandwidth of a typical video file. Assume the file has a resolution of 320x240, which means that there are 76,800 pixels in each frame. Each raw pixel represents one byte of data.

We captured the video at 24-bits color depth, so it takes 24 bits, or 3 bytes, of data to describe the color of every pixel. As we discussed, the high color depth is necessary to simulate all of the colors in the original analog video. However, it also means that each video frame, uncompressed, is about 230 kB in size.

We captured the video at 15 frames per second. That translates to about 3.456 MB of data per second, or over 207 MB per minute. In one hour you'd accumulate over 12 gigabytes of data. Probably a bit more than you have on your hard drive, wouldn't you say?

	Per Frame	Per Second	Per Minute	Per Hour
Uncompressed Bandwidth	230kB	3.456 MB	207 MB	12.41 GB

It also exceeds the transfer bandwidth of most computer system peripherals. Double spin CD-ROM drives transfer about 300 kB/s, less than a tenth of what's required. ISA bus computers can transfer about 2.5 MB/s through the bus, about two-thirds of what's required. So even after scaling video resolution from 512x480 to 320x240, and dropping the frame rate from 30 to 15, our uncompressed video is still too large to store economically, and impossible to access in real time.

That's where video compression comes in. For without compression, there would be no digital video.

SUMMARY

1. Digital video is video playing on your computer in digital format, not analog video playing through a special adapter card.

2. Most video is originally filmed in analog format, a naturally continuous signal with breadth and depth. When video is digitized, or converted to digital format, precise color values are assigned to fixed points, or pixels, in the analog stream. This allows the digital computer to represent an essentially analog event.

3. Digitization requires both video and audio capture facilities and an analog source to feed the video stream into the computer. During capture, video is reduced in frame rate and resolution, or scaled. Capture software like VidCap, the Video for Windows program, controls the process, including synchronization of the audio and video streams.

4. Bandwidth has two meanings. The first is a measure of a device's capacity or ability to move information to or from a device, system, or subsystem. For example, a single spin CD-ROM drive can only transfer 150 kB/s into the computer. Accordingly, the CD-ROM's bandwidth is 150 kB/s.

    The second meaning refers to the data flow necessary to support a process like video playback. For example, a video file that's 30 seconds long and approximately 4.5 MB in size has an average bandwidth of approximately 150 kB/s, meaning that you couldn't play this video from a device with a bandwidth capacity of less than 150 kB/s. When used in this manner, bandwidth is used interchangably with the term data rate.

5. Even after scaling to a smaller window size and cutting the frame rate from 30 to 15, raw video has a bandwidth of 3.456 MB/s. This exceeds the bandwidth of all but the fastest computer storage devices and is impossible to store economically. For this reason, compression is essential to digital video.

INTRODUCTION TO VIDEO COMPRESSION

2

Introduction to Video Compression

Video compression is the science of the impossible. As we've seen, without compression, most computers can't supply the data, much less decompress and display at 15 frames a second.

The currency of the impossible is the trade-off. Technologies trade off screen resolution for file size, color depth for frame rate and video quality for data rate—all to deliver a steadily improving but still less than dazzling video stream.

This section starts by defining compression, then moves to the decompression and display process. This will help you appreciate how much work goes into playing digital video files. Then we'll study the two main elements of video compression, interframe and intraframe compression.

One of the most often-asked questions is why cheap television sets outperform Pentium computer systems respecting video playback—in essence, why analog video is superior to its digital counterpart. We'll address that herein.

Finally, the classic compression trade-off is video quality for bandwidth, and as file size approaches CD-ROM rates some loss in quality is inevitable. However, video compression works extremely well in some cases and very poorly in others. Understanding the compression mechanics presented in this chapter will help you work with compression—not against it—and help you produce higher-quality compressed video.

A WORKING DEFINITION OF COMPRESSION

Figure 2.1 Definition of compression

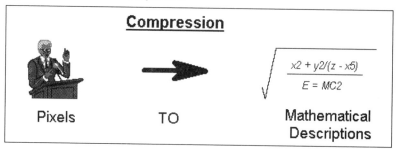

As we've seen, uncompressed video is a sequence of frames containing pixels. Video compression is a process where a collection of algorithms and techniques replace the original pixel-related information with more compact mathematical descriptions.

Decompression is the reverse process of decoding the mathematical descriptions back to pixels for ultimate display. At its best, video compression is transparent, or invisible to the end user. Video consumers—the end users actually watching the video—don't want compression, they want video. So the true measure of a compression technology is how little you notice its presence, or how effectively it can reduce video data rates without adversely effecting video quality.

Figure 2.2 Definition of decompression

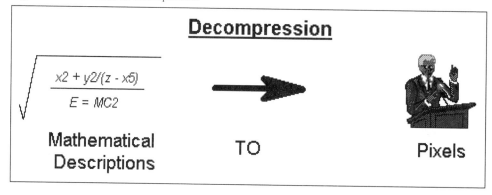

Compression Fundamentals

Video compression utilizes two basic compression techniques: interframe compression, or compression between frames, and intraframe compression, which occurs within individual frames. We'll cover both topics in detail in a few moments.

A few other points. First, the techniques and algorithms used during compression transform a video frame from a collection of dumb pixels to a set of instructions used by the decompressor to recreate the original pixels. So when we describe a frame "telling" the decompressor what to do, you'll know what we mean.

Second, compression doesn't change the number of frames in the video. During scaling, you may drop the frame rate from 30 frames per second to 15. However, once you submit a video to the compressor, the frame count doesn't change. Even during interframe compression, where frames "borrow" information from other frames, each frame contains its own separate description.

Third, while all products use intraframe compression, not all use interframe. For example, Xing Technologies' Scalable MPEG does not use interframe techniques, while most other technologies do.

Finally, you can't buy interframe compression from one vendor, and intraframe from another. You buy one product from one vendor that takes in digitized video and produces compressed files. The same with decompression, which is always dedicated to one compression technology. This compression/decompression combo is commonly referred to as a *codec*.

Lossless vs. Lossy Compression

One of the most fundamental concepts in compression is the difference between lossless and lossy compression. Lossless compression techniques create compressed files that decompress into *exactly* the same file as the original, bit for bit. Lossless compression is typically used for EXE and data files where any change in digital make-up renders the file useless.

Lossless compression is used by products such as STAC and DoubleSpace to transparently expand hard drive capacity, and by products like PKZIP to pack more data onto floppy drives for delivery or storage.

In general terms, lossless compression techniques identify and utilize patterns within files to describe the content more efficiently. This works quite well for files with significant redundancy, such as database or spreadsheet files. As a whole, however, these techniques typically yield only about 2:1 compression, which barely dents uncompressed video files.

Lossy compression, used primarily on still image and video files, creates compressed files that decompress into images that look similar to the original but are different in digital make-up. This "loss" enables such techniques to deliver from 30:1 to 50:1 compression.

Run an EXE file through a lossy compressor and you've got serious problems. However, in a 24-bit image file, a few changed pixels or altered shades is virtually unnoticeable. This is especially true in video, where your eye has to spot defects in 1/15 of a second.

Lossy Compression—Example

Figure 2.3 is a 180 Kilobyte TIF file. Figure 2.4 was compressed to about 4 kB using a lossy compression technique called JPEG. At about 45:1 compression, JPEG was over 22 times more effective than most lossless techniques. Even upon close scrutiny, it's almost impossible to tell the images apart.

You might be thinking—4 kB per frame, 15 fps, that's a bandwidth of 60 kB/s and the video looks great. What's the problem? We'll just use JPEG.

Well, unfortunately, it doesn't work that way. JPEG implementations can decompress and display a file very quickly, often in under a second on 80486 class machines. However, video requires *15 frames per second.* While still image codecs can focus on quality at the lowest possible file size, video codecs must be concerned with quality and *display rate*, or the rate at which frames decompress and display. Otherwise, they

Figure 2.3 Original file—180 kB

Figure 2.4 JPEG file—4 kB

would produce video that looked like fast slide shows with audio.

For this reason, video codecs aren't the sequential operation of lossy still-image codecs; they're built from the start to provide the optimal balance among display rate, video quality and bandwidth.

All relevant codecs are lossy technologies—it's the only way to achieve the required compression. By both definition and operation, this translates to loss in quality, which is the first reason that digital video doesn't look as good as analog video. The second reason relates to the cumbersome nature of the decompression and display process. That's our next stop.

Decompression and Display

Video decompression and display takes five discrete steps. Understanding this process helps you appreciate why video chokes most computers. It's also critical to understanding subtle differences between codecs and how innovations such as local bus and video co-processing accelerate video performance.

To illustrate this process, we've recruited some real specialists. Our first helper is Mikey—who represents your computers' host processor. Since we're running in Windows, that means he's a 386, 486 or Pentium. Mikey's a painter—which is pretty appropriate given the amount of painting done during video playback.

Vinnie represents your video processor. As you probably know, every video card has its own processor. Vinnie's role in digital video has traditionally been very small, but it's growing rapidly, and he'll be a major player in the near future.

You'll meet Buster, a bus coprocessor, later. A relative new comer to the computer arena, Buster also plays an important role in digital video's future.

 Video—You can view a live demonstration of this process by double-clicking on PBACKBAS.AVI in the chap_2 subdirectory.

Figure 2.5 Step 1—Data retrieval

Retrieval—The first step is retrieving the compressed data stream into main memory, which is typically performed by the host central processing unit, or CPU. Here the data is being retrieved from a CD-ROM. As shown in Table 2.1, this simple transfer requires a surprising amount of processor overhead.

Note that retrieval from a CD-ROM takes almost twice the CPU overhead. As

Table 2.1 CPU overhead required to retrieve data at 150 kB/second. Tests performed on 80486 VLB bus computer with 8 MB RAM

	Hard Drive	CD-ROM
150kB/s transfer	11.1%	21.10%

Figure 2.6 Step 2—decompression

we'll see, that can mean that video plays faster from a hard drive than from a CD-ROM.

Decompression—After the video is main memory, the CPU converts the mathematical descriptions contained in the frames back to pixels to start the display process.

Transfer to Windows Buffer—After decompression, the data must be converted to Windows Device Independent Bitmaps (DIBs) and placed in a memory buffer for transfer to the graphics board. We're showing this as painting to a Windows buffer.

Figure 2.7 Step 3—paint to a Windows buffer

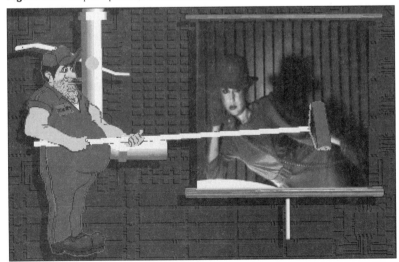

Figure 2.8 Windows Graphics Device Interface

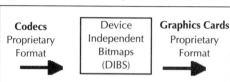

In Windows, codecs can't communicate directly to video hardware; they operate through Windows Graphics Device Interface, or GDI. This is the structure defined by Microsoft to establish compatibility between various hardware and software peripherals (Figure 2.8).

All codecs store compressed video data in proprietary formats. Similarly, all graphics cards use proprietary data structures to work with video data. Windows GDI established the DIBs as the "common language" spoken by these peripherals, allowing different products to work together without writing specific device drivers.

This conversion and associated overhead is the reason that many Windows programs are slower than their DOS counterparts. For most applications, however, the overall effect isn't significant. For example, a half of a second delay in displaying an image or reformatting a paragraph isn't a big deal. In video applications, where 15 frames per second is *de rigeur*, Windows GDI is a significant burden which costs several frames per second during decompression and display.

Transfer to Video Memory—The fourth step is another bus transfer from main memory to the video card, also performed by the host CPU (Figure 2.9). At this point, the video is fully decompressed, so our once-compact data stream is now back at several megabytes per second. This transfer staggers many local bus computers and brings most ISA bus computers to their knees. However, this bus transfer is the final step for the host CPU, freeing it to retrieve another frame and start the decompression and display process again.

Figure 2.9 Step 3—transfer to video memory

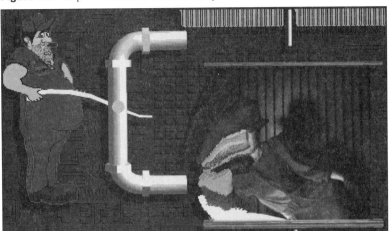

Paint to Video Memory—At this point the video processor takes over, painting the video to video memory and converting the data back to analog format for display.

This time- and CPU cycle–consuming process all relates back to the fact that video is an analog format that digital

Figure 2.10 Step 5—paint to video memory

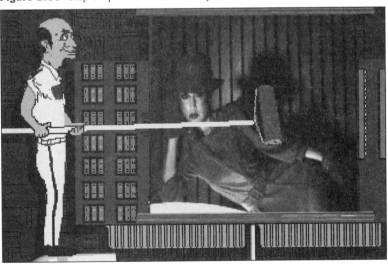

computers can only simulate, never operate in its native form. In contrast, your television takes in an analog signal and simply transfers it to the picture tube, which quickly paints lines down the screen. No decompressions, constricted bus transfers or file conversions. Small wonder a $200 television can outperform your 486 when it comes to video.

INTERFRAME COMPRESSION

Interframe compression uses a system of key and delta frames to eliminate redundant information between frames. Key frames store an entire frame, while delta, or difference, frames record only interframe changes.

Key frames, or reference frames, are not compressed at this stage. Instead, they're included in the compressed stream in their entirety to serve as a reference source for delta frames. Delta frames contain only pixels that are different from either the key frame or the immediately preceding delta frame, whichever frame they reference during compression.

During decompression, delta frames look back to their respective reference frame to fill in missing information.

Here's how it works mechanically. Figure 2.11, or frame 13857, is the key frame, while frame 13859, shown in Figure 2.12, is the delta frame. The frames are two apart because the video sequence from which they were taken was captured and compressed at 15 frames per second. Both frames are divided into blocks of pixels. The delta frames are compared, block by block, with their respective key frames. Blocks that match are discarded. Blocks that don't match are saved, as shown in Figure 2.13, and subsequently compressed during intraframe compression.

During decompression, key frames are carried forward to help construct delta frames. In essence, during decompression the delta frame tells the codec "I'm just like the key frame except for these pixels." Or, in our example, change the numbers, adjust for some minor face movement and use the rest from the key frame. The decompressor makes the changes and converts the frame to a Windows DIB for transfer to the video card.

Figure 2.11 The key frame

Figure 2.12 The delta frame

Figure 2.13 Delta frame with interframe redundancies removed

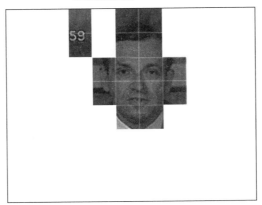

Video—You can view a live demonstration of this process by double-clicking on INTERFRA.AVI in the chap_2 subdirectory.

Interframe Variants

Different compression techniques use different sequences of key and delta frames. Most Video for Windows codecs, for

Figure 2.14 Delta/Key frame sequence for typical Video for Windows codec

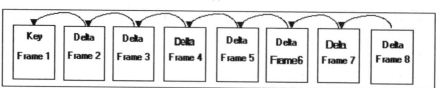

example, calculate interframe differences between sequential delta frames during compression. In this scheme, shown in Figure 2.14, only the first delta frame relates back to the key frame; the rest relate back to the immediately preceding delta frame. In other schemes, all delta frames relate back to the key frame.

Finally, MPEG, a hardware-based compression technology, uses three kinds of frames: an I or Intra Frame, which corresponds to the key frame; a P, or predictive frame, which roughly corresponds to the delta frame; and a B, or bi-directional frame, which is a delta frame that uses both past and future frames as reference sources. More on MPEG later.

Whatever the scheme, all intraframe compression techniques derive their effectiveness from interframe redundancy. Low-motion sequences like the "talking head" shown in Figure 2.11 have a high degree of interframe redundancy, which limits the amount of *intraframe* compression required to reduce the video to the target bandwidth.

All relevant intraframe techniques are lossy in nature. When there's less data to compress there's less loss in quality. During playback of low-motion sequences, less decompression is required, which reduces the load on the host CPU and boosts the display rate, increasing perceived video quality.

Interframe motion is the enemy of interframe compression, decreasing both video quality and display rate. Let's see just how badly.

Figure 2.15 is a raw frame from a high motion sequence. Figure 2.16 is the same frame, compressed to 150 kB/s with Video 1, the same codec that produced the high-quality talking head image shown earlier in the chapter. As you can see, quality dropped dramatically and the motion video is practically unusable. Not only did quality suffer, the display rate for the second video was 3 frames per second slower than the first.

Figure 2.15 High-motion sequence, uncompressed

Figure 2.16 Same frame, compressed

As you may have guessed, Video 1 is not our first choice for compressing high-motion sequences. However, while the effects may not be as severe, motion degrades the video quality of all codecs.

Dynamic Carryforwards

So far, in calculating interframe compression, we've only looked at pixel blocks that remained static between the delta and reference frame. Some codecs, however, increase compression by tracking moving blocks of pixels from frame to frame. This is called motion compensation, or dynamic carryforwards, since the data carried forward from key frames is dynamic, not static.

If you tracked only static pixels between the frames shown in Figures 2.17 and 2.18, no interframe compression would occur with respect to the arm, because it's not located in the same pixel blocks in both frames. However, if the codec could track the motion of the arm, the delta frame description could tell the decompressor to "look for the arm in the key frame and move it over four blocks to the right."

In operation, these techniques are implemented block by block. For example, when compressing delta frame pixel block J-9, the codec would first search for an exact match in pixel block J-9 of the key frame. It wouldn't find a match. Then it would search within the range prescribed by the codec. Codecs

Figure 2.17 Key frame

Figure 2.17 Delta Frame

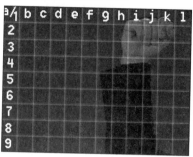

designed to maximize compression speed might limit searching to one or two pixel blocks in either direction. Codecs designed to maximize video quality might search in a much greater range.

If pixel block F-9 was in that search range, the codec would note the location and save the motion vector, or directions for getting to the key frame block, in the delta frame description. In this case the motion vector would be four blocks to the right.

During decompression, the delta frame essentially tells the codec to use the information contained in pixel block F-9, but move it over four blocks. Since this is more efficient than compressing the entire arm with intraframe techniques, additional compression results.

Video—You can view a live demonstration of this process by double-clicking on DYNAMICC.AVI in the chap_2 subdirectory.

Even though dynamic carryforwards are universally helpful, they are not universally applied. For example, real time capture boards typically can't scale resolution and frame rate, digitize *and* hunt for dynamic carryforwards at the same time—capture alone is too intense.

Similarly, tracking motion vectors during decompression is a processor-intensive task, almost like assembling 15 puzzles a second. This is much more difficult than static carryforwards. Most codecs can't implement dynamic carryforwards without slowing video display rate to a crawl. So most technologies preserve decompression speed by limiting the number of vectors tracked or don't implement dynamic carryforwards at all.

As a general rule, techniques such as dynamic carryforwards mark the point of demarcation between software-only codecs and hardware schemes such as MPEG and DVI. Software-only codecs such as Indeo and Cinepak decompress without special hardware, with performance typically limited by processor and bus type. They can't implement advanced tracking schemes and maintain a reasonable display rate.

Hardware schemes rely on dedicated playback boards for the horsepower to decompress dynamically tracked pixel blocks, to assemble the puzzle at 30 frames per second. This allows them to produce a generally higher level of video quality and to display 30 frames per second across a wide range of target computers.

Software schemes offer widespread compatibility and low-cost playback. Hardware schemes offer consistent performance at the price of dedicated hardware on the target stations.

As computers get faster, the lines between hardware and software begin to blur. We'll cover this in detail in subsequent chapters, but the point to remember is this. Hardware systems don't offer magically advanced technology. In many ways, the software codecs are much more sophisticated. Hardware technologies simply free the codec to use well-known techniques such as dynamic carryforwards to compress to higher quality while maintaining 30 fps display rate during playback.

Putting Interframe Theory to Work

As we've seen, the degree of interframe compression significantly affects how well a video sequence will compress. If your video doesn't compress well, it won't look good, especially when compressing to a fixed bandwidth.

During the development of popular multimedia titles such as Sherlock Holmes from ICOM Simulations, a lot of energy went into creating compressible video sequences. For example, while shooting Sherlock Holmes, the developer eliminated camera motion altogether, and coached the actors to minimize their own movements.

Since Sherlock Holmes was published in 1992, codec technology has advanced significantly. However, the lesson still holds. When filming for compression, you should plan your shoot to minimize incidental motion. This lets you maximize the quality of motion that is critical to delivering the intended content of the video.

Many common analog film techniques cripple interframe compression. For example, panning, or moving the camera slowly across a field or object, changes every pixel every frame. Similarly, zooming, either closer to or further away from the subject also changes every pixel every frame.

Filming without a tripod often introduces motion that's imperceptible in the analog world but extremely damaging to digital video. Prior to one COMDEX in Atlanta, I borrowed some footage from a buddy who worked with Deion Sanders' show on TBS. One sequence that looked great on the VCR was Deion interviewing Barry Bonds. However, the scene compressed very poorly. When we closely examined the original footage, we noticed that the video shifted one or two pixels up and down virtually every frame, and later learned that the show was shot with a handheld camera.

Had the video been shot with a tripod, our COMDEX demonstration would have been spectacular, and who knows, maybe the Braves wouldn't have traded Deion to Cincinnati. Instead, the footage was unusable.

It's also useful to select sequences with clean foreground and background patterns. All analog video has noise or distortion which is increased during capture and compression. For example, if you look closely at your television set, you'll notice that the background typically shimmers slightly almost all the time. Sit in the front row at your local theater and you can see the objects move what looks like a few inches between frames, especially text like the credits. This noise can transform a striped shirt into a blotchy solid and a complex background pattern into a boiling mass of colors and motion.

I recently spoke before a group of computer based training executives at a seminar in San Jose. After briefly describing the points I just listed, one attendee raised his hand and said, "You're eliminating all the fun effects that make video interesting—which is why we use video in the first place. If all we wanted to use was static talking heads, who needs it?"

His question came very late in the seminar and caught me off guard so I didn't really have a good answer. But after some reflection, here's what I think.

The state of the art in digital video is the state of the art—it's what it is. Several codecs can compress high-motion sequences and deliver good quality at reasonable data rates, so most video shots are possible. However, interframe motion stresses compression—it's a fact. And arty pans and zooms and other extraneous motion will detract from video quality.

The best approach is to analyze which motion is critical to delivering the intended message of the video. Then try to minimize all others. This lets you deliver your true message with the highest possible quality video.

It also goes without saying that proper lighting and similar techniques are also critical to capturing high-quality analog footage. What makes multimedia so difficult is that most of the tasks are new—you not only have to have a great idea, you have to be a script writer, cameraman, director, editor, creative director and often the star as well. Then you have to digitize it, code it, package it and sell it.

This is a long way of saying two things. First, there's not much I can add about proper filming techniques except what's just been presented. Second, if you're going to invest tens of thousands of dollars in a project, you probably should invest in a professional photographer and director—at least the first time out.

Interframe Summary

A few closing comments about interframe compression. First, when comparing codecs, remember that even minor differences in interframe motion can dramatically affect performance. So when you look at demos, be sure to notice the amount of motion. Better yet, use identical test clips. Finally, never judge a technology solely by its talking head performance, especially if you'll be working with moderate action scenes. As we saw in Figure 2.8, the ability to produce high-quality low-motion sequences doesn't automatically translate into the ability to produce high-quality action sequences.

INTRAFRAME COMPRESSION

Intraframe compression is performed solely with reference to information within a frame. It's performed on pixels in delta frames that remain after interframe compression, and on key frames.

While intraframe techniques are often the most hyped, overall codec performance relates more to interframe efficiency than intraframe. For this reason, we'll provide only brief descriptions of the intraframe technologies and focus the bulk of our attention on the individual codecs.

Run Length Encoding

Run length encoding, or RLE, is a simple lossless technique originally designed for data compression and later modified for facsimile. RLE essentially encodes "runs" of different pixel lengths.

Imagine this page was divided into 2,200 lines and 1,700 columns, or 200 dots per inch for the 9.5 inch height and 7.25 inch width. This is the maximum resolution of Group III facsimile. Look at the blank line immediately above this paragraph. You could describe the line as 1700 discrete white pixels, which would take about 1,700 bytes of data. Alternatively, using RLE, you could define the pixels as one run of 1,700 white pixels, which would be much more efficient.

Most 24-bit real-world videos, however, don't have long runs of identically colored pixels. While RLE works well in the black-and-white facsimile world, it's an extremely inefficient mechanism for video.

JPEG

JPEG stands for the Joint Photographic Experts Group. JPEG has been adopted as a standard by two international standards organizations, the CCITT and the ISO, and is the most prevalent still image compression technology. JPEG operates through the following three-step process.

Step one is the encoding mode, where the data is converted into frequency space for the discrete cosign transform (DCT) analysis. DCT starts by dividing the image into 8x8 blocks like those shown in Figure 2.19, then converts the colors and pixels into frequency space by describing each block in terms of the number of color shifts (frequency) and the extent of the change (amplitudes).

Because most natural images are relatively smooth, the changes that occur most often, or high-frequency changes, have low amplitude values, meaning that the change is minor. In other words, images have many subtle shifts among similar colors, but few dramatic shifts between very different colors.

Figure 2.19 JPEG breaks image into blocks of 8x8 pixels

In the next JPEG stage, quantization, amplitude values are categorized by frequency and averaged. This is the lossy stage because the original values are permanently discarded. However, because most of the picture is categorized in the high-frequency/low-amplitude range, most of the "loss" occurs among subtle shifts that are largely indiscernible to the human eye. This concentrates the bulk of the compressed file information on low-frequency/high-amplitude changes such as edges and corners.

For example, Figure 2.19 has many shades of gray which are the high frequency/low amplitude changes. If the grays blended together, your eye probably wouldn't notice. The highest-amplitude change is between the gray suit and white shirt. If this was blended into a mottled gray, it would be extremely noticeable because the edge would be lost and two colors dramatically shifted. JPEG's quantization mechanism avoids this by focusing most of the loss where it won't be noticed.

After quantization, the values are further compressed through RLE using a special zigzag pattern designed to optimize the compression of like regions within the image.

At extremely high compression ratios, more high-frequency/ low-amplitude changes get averaged, which can cause an entire pixel block to adopt the same color. This cause a blockiness that's characteristic of JPEG-compressed images.

JPEG's relatively simple mechanics make it extremely fast. It is also a symmetric algorithm, meaning that decompression is exactly the opposite process of compression and occurs just as fast. JPEG is used as the intraframe technique for MPEG.

Vector Quantization

Like JPEG, Vector Quantization, or VQ, also divides the images into 8x8 blocks, but the information quantized is completely different. VQ is a recursive, or multistep, algorithm, with inherently self-correcting features. Here's how it works.

The first step is separating similar blocks into categories and building a "reference" block for each category. The original blocks are all discarded. During decompression, the single reference block will replace all of the original blocks in the category.

After selecting the first set of reference blocks, you decompress the image and compare it to the original. Typically there will be many differences, so you create an additional set of reference blocks that fill in the gaps created during the first estimation. This is the self-correcting aspect of the algorithm. Then you repeat the process to find a third set of reference blocks to fill in the remaining gaps—more self-correction. These reference blocks are all posted in a look-up table to be used during decompression. The final step is to use lossless techniques such as RLE to further compress the remaining information.

VQ compression is obviously computationally intensive. However, decompression, which simply involves pulling values from the look-up table, is extremely simple and fast. VQ is a public-domain algorithm used as the intraframe technique for both Cinepak and Indeo.

Wavelets

Wavelets combine completely different method of image analysis with the best features of JPEG and VQ. The first step is to filter the image with a high-pass and low-pass filter, essentially creating multiple views of the image. This clearly identifies the location of low-frequency/high-amplitude information and high-frequency/low-amplitude information. A tree-based encoder works through the different views of the image, starting with low-frequency data.

After each encoding run, the codec checks the encoded information against the next image view. This is similar to the self-correcting aspects of VQ. If the higher-frequency information doesn't accurately describe the next view, the difference data is encoded. Otherwise, no additional information is encoded and the codec begins searching the next level, checking for accuracy at each level.

Once the image is completely encoded, a simple quantization method similar to JPEG's is used to compress the data.

Once again, the high-frequency/low-amplitude information is compressed more than the low-frequency/high-amplitude information. This preserves edges and targets the bulk of the compression towards areas not readily noticed by the human eye.

The final step is lossless compression through Huffman Encoding, an advanced variant of RLE.

Wavelets is a symmetric technology that compresses and decompresses very quickly. A public-domain algorithm, Wavelets are the interframe technology implemented in Media Vision's widely heralded but short lived Captain Crunch codec.

Fractal Compression

Fractal compression results from a patented process invented by Dr. Michael Barnsley, founder and chief technical officer of Iterated Systems. Like JPEG and VQ, fractal compression starts by breaking the image into blocks. The next step, similar to VQ, is comparing blocks to other larger blocks to find similar blocks. However, rather than storing this information in a look-up table, the fractal process converts these self-similar regions into equations that can be used to recreate the image. In essence, the process converts the original pixel-related information into a mathematical model of the image.

This mathematical model is technically resolution-independent, meaning that fractal images can be zoomed to any resolution, irrespective of the original size. This characteristic, called scalability, is the primary advantage of fractal techniques.

Searching for self-similar regions is a lengthy process, and fractal compression is extremely time-consuming. However, the technology is asymmetric, and decompression is extremely simple and fast. Fractal compression is the intraframe technology used by Iterated Systems for their fractal video codecs.

SUMMARY

1. Video compression is a process where a collection of algorithms and techniques replace the original pixel-related information with more compact mathematical descriptions. Decompression is the reverse process of decoding the mathematical descriptions back to pixels for ultimate display.

2. Video compression utilizes two basic compression techniques: interframe compression, or compression between frames; and intraframe compression, which occurs within individual frames.

3. Lossless compression techniques create compressed files that decompress into exactly the same file as the original, bit for bit. Lossy compression techniques create files that decompress into files that are similar to the original, but are different in digital makeup.

   All relevant video codecs are lossy in nature—it's the only way to achieve the required compression. This sets up the classic compression trade-off, video quality for file size. It's also the first reason why digital video doesn't look as good as analog video.

4. Decompression and Display is a five step process:

   (a) Retrieval from storage and transfer to RAM.

   (b) Decompression

   (c) Painting to Window buffer

   (d) Transfer to video card

   (e) Paint in video memory

   In contrast, your television simply receives the analog data, transfers it to the tube and paints scan lines down the screen. No conversions, decompressions or restricted bus transfers. No wonder a $200 television looks better than your Pentium when it comes to video.

5. Interframe compression uses a system of key and delta frames to eliminate redundant information between frames. Key frames are reference frames for delta frames and are not compressed during interframe compression. Delta frames contain only pixels that are different from the reference frame.

6. All interframe techniques derive their effectiveness from interframe redundancy, or the absence of motion. Videos with low motion content compress more effectively than high motion videos, which means better video quality at the same bandwidth.

7. Dyanamic Carryforwards, or Motion Compensation, are techniques used to expand interframe compression by tracking moving pixels between the frames. However, these techniques also slow decompression speed and are therefore used sparingly by most software-only codecs.

8. Plan your video shoots to minimize extraneous motion. Avoid panning, zooming and frequent cuts where possible, and always file with a tripod. The best approach is to analyze which motion is critical to delivering the intended message of the video. Then try to minimize all others. This lets you deliver your true message with the highest possible quality video.

9. Intraframe compression is compression performed solely with reference to an individual frame. It's performed on pixels in delta frames that remain after interframe compression and on key frames. While intraframe compression techniques are the most hyped, usually codec performance relates more to interframe techniques than intraframe.

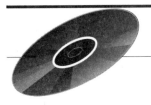

INTRODUCTION TO VIDEO FOR WINDOWS

3

INTRODUCTION TO VIDEO FOR WINDOWS

IN THIS CHAPTER

In large part, video under Windows is defined by Video for Windows, an architecture and application suite introduced by Microsoft in November 1992. The bulk of our work later in the book relates to tools and codecs included with or spawned by the Microsoft specification.

This chapter describes Video for Windows' architecture and presents an overview of the various Video for Windows components and codecs.

OVERVIEW

Before Video for Windows, the video market was composed of many separate and incompatible products. All codecs used proprietary programming interfaces that complicated video creation and integration. Most capture boards were codec-specific, and there were few mechanisms to convert to and from the different codecs. This scenario was fine for Intel and IBM, but a problem for small codec vendors without the expertise or financing to build their own hardware.

Multimedia market growth on the Windows platform was retarded by the lack of standards that were in full bloom on the Macintosh platform with Apple's QuickTime standard. This was the problem Microsoft sought to address with Video for Windows, first introduced in November 1992 as a retail product priced at $199.

Figure 3.1 Video for Windows components

```
┌─────────────────────────────────────┐
│         Video For Windows           │
│                                     │
│  ◆  Architecture                    │
│                                     │
│        ◆  Technology IN             │
│        ◆  Multimedia OUT            │
│                                     │
│  ◆  Compression Application Suite   │
└─────────────────────────────────────┘
```

Video for Windows is both an architecture and an application suite (Fig. 3.1). As an architecture, Video for Windows provides both inbound and outbound interfaces. The inbound architecture ensures that various multimedia technologies, including codecs, work together to form a comprehensive multimedia creation and playback platform. By defining a standard file format, called the Audio/Video Interleaved (AVI) format, Microsoft opened the market for capture boards built by third-party hardware developers. By supporting the AVI file format, Video for Windows codec developers can access video captured by these boards automatically. This architecture provides equal access to peripherals and lets individual developers focus on their key technology.

The AVI file format also opened up markets for other peripheral products such as video editing software—for example, Adobe Premiere and MediaMerge from ATI Technologies. The AVI standard helps Adobe and ATI by creating a market with sufficient critical mass to profitably support this type of product. Similarly, all Video for Windows' codecs can claim access to these essential products simply by supporting the AVI format.

As an outbound architecture, Video for Windows provides two outbound APIs, discussed in detail later, for integrating video into other applications. As an application suite, Video for Windows primarily handles video capture and compression, with some ancillary video and audio editing functions.

During 1993, Video for Windows was widely criticized for selling "only" 50,000 copies, most bundled with capture boards. But an April 1994 *PC Magazine* review I worked on analyzed seven capture boards that cost under $700. In April of 1993, there were no boards in this price range. Similarly, the review tested five video editors; the previous year there were none. Clearly, Video for Windows was a success as measured by how much it helped ease access to video on the Windows platform.

ARCHITECTURE

Figure 3.2 Dual components of the VIDEO FOR WINDOWS playback

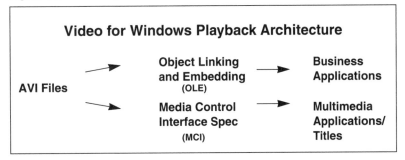

Outbound APIs

Video for Windows has two outbound Application Programming Interfaces (Fig 3.2), or APIs. The most comprehensive is a set of extensions to the Media Control Interface (MCI), an API for communicating with multimedia devices jointly released by Microsoft and IBM in 1991. MCI controls are similar to Basic commands, and the interface was primarily designed for use by programmers.

Technically, Video for Windows is a device under the MCI commands. This means that while all Video for Windows commands are accessible through the MCI interface, the reverse is not true. As we'll see, some devices and codecs are MCI-compatible, but not Video for Windows–compatible.

Media Player, an application shipped as an accessory program with all versions of Windows, provides a more accessible video playback interface. Media Player is essentially an MCI device that recognizes and plays most MCI output formats, including Video for Windows' AVI files.

Figure 3.3 Media Player, showing option to zoom by 2 and File
Information box

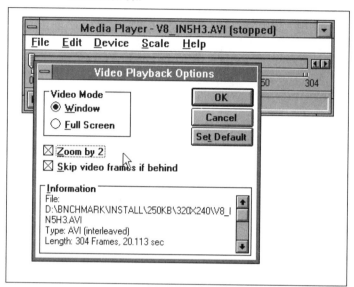

Media Player is also a server under Microsoft's Object
Linking and Embedding (OLE) specification. This means that
Media Player can link into and play back supported multime-
dia elements into OLE Client applications such as Lotus Ami
Pro and Microsoft Excel.

While Media Player is easier to use, its capabilities are limited. For example, as shown in Figure 3.3, Media Player can
play video in only two resolutions, single- and double-scale. In
contrast, MCI controls enable playback in virtually any sized
window. This type of flexibility makes the MCI specification
the interface of choice for applications developers and multimedia publishers.

MCI Application Support

Many authoring and presentation programs have opted to
build MCI support directly into their products, providing a

third level of access to Video for Windows. Users can incorporate video files into such programs through controls *in that program,* instead of OLE or MCI. This simplifies the integration process and provides an intermediate level of flexibility of use.

For example, Asymetrix Compel, used to prototype Doceo's *Video Compression Guide and Toolkit,* directly supports many multimedia formats, including digital video (see Figure 3.4). Video, audio and animation were incorporated into the prototype with Compel controls, which was faster and easier than either MCI or OLE.

In contrast, some authoring programs and most programming languages force you to add video and other multimedia elements through the actual MCI commands. This is often much harder from a programming standpoint, but affords much greater control and flexibility than through typical means of direct support.

The clear trend is towards integrating MCI commands directly into applications and authoring languages, which effectively shields the end user from both MCI and OLE. This will simplify the use of MCI-controlled media elements, including AVI files, and thereby promote the multimedia usage and overall market growth.

Figure 3.4 AVI files integrated into Compel presentation through direct MCI support

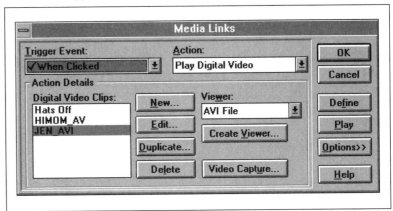

Codec Architecture

On the inbound side, Video for Windows provides a standard interface for multimedia tools and subsystems, including codecs. Microsoft has licensed or developed four codecs to ship with Video for Windows: Cinepak from SuperMac, Intel's Indeo, Microsoft's RLE and Microsoft /Media Vision's Video 1 (see Figure 3.5). As we'll see later, each codec has its own strengths and weaknesses.

While all Video for Windows codecs produce files with the AVI extension, the files are all different. For this reason, each codec has a dedicated playback module that must be installed to play the video.

Distribution of video files created with codecs bundled with Video for Windows is royalty-free. This means that you can

Figure 3.5 VidEdit's Compression Options screen, showing four Video for Windows codecs and Captain Crunch

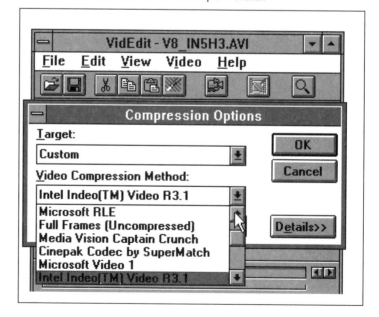

compress a video file with any codec and distribute the file internally or commercially without royalty. You can also distribute the dedicated playback module, or runtime, royalty-free, as we have with the bundled CD-ROM.

Other Codecs

Not all Windows-based codecs are compatible with Video for Windows. For example, Apple Computer's QuickTime for Windows and versions of Xing Technologies' Scalable MPEG are not Video for Windows–compatible. Both companies have valid reasons for not supporting the specification. Apple is promoting QuickTime as an alternative multimedia standard, while Xing reports performance degradation under Video for Windows.

These codecs can't directly use many Video for Windows tools and peripherals, which makes working with these format less convenient. While Apple has garnered QuickTime support from many developers of authoring and presentation programs, Xing hasn't. As a result, you can only edit Xing video files with Xing's video editor.

While these formats are incompatible with Video for Windows, they do comply with the deeper MCI specification, and developers can access the codecs through MCI commands. Since Media Player is an MCI device, it can recognize and play back MPEG and QuickTime files with the proper drivers. End users can then access these codecs through Media Player's OLE server capabilities.

In sum, Video for Windows compatibility has many benefits, including ease of access to capture, editing, compression and playback programs. However, not all codecs are Video for Windows–compatible. When evaluating codecs, Video for Windows compatibility should be an important criterion— otherwise you could be forced to work with a much more limited set of tools than those available for Video for Windows codecs.

VIDEO FOR WINDOWS APPLICATIONS

Figure 3.6 Video for Windows' capture program, VidCap

VidCap

Two programs, VidCap and VidEdit, provide the bulk of Video for Windows' applications capabilities. VidCap is the primary capture application and provides the interface between analog sources such as laserdiscs and video cameras and video and audio capture boards (Figure 3.6). VidCap can capture still images or video sequences, with or without audio.

VidCap can capture video in either compressed or raw format, depending on the capabilities of the capture card. Either way, it produces an AVI file with synchronized video and audio.

VidCap used to be the only Video for Windows capture application. In Video for Windows 1.1, released November 1993, however, Microsoft provided a comprehensive API to the capture functions embodied in VidCap and VidEdit. These APIs let other applications incorporate these capabilities through their own application interface.

Before these APIs were available, applications could capture video, but only by calling VidCap. For example, when you select video capture in Adobe Premiere version 1.0, the program loads VidCap to capture the file. This presents the user with two different program interfaces, which can be confusing. In addition, since VidCap is only available in Video for

Windows, it forces the user to buy or otherwise acquire Video for Windows.

In Premiere version 1.1, Adobe implemented these drivers and captures video through its own interface. VidCap is no longer required. This integrates capture functionality more cleanly from a user-interface perspective and eliminates the need to acquire VidCap. It also frees Adobe to advance video capture functionality in ways that were impossible when VidCap was required for capture.

VidEdit

VidEdit is an AVI editor and compression manager for captured video files (Fig 3.7). With VidEdit you can scale to

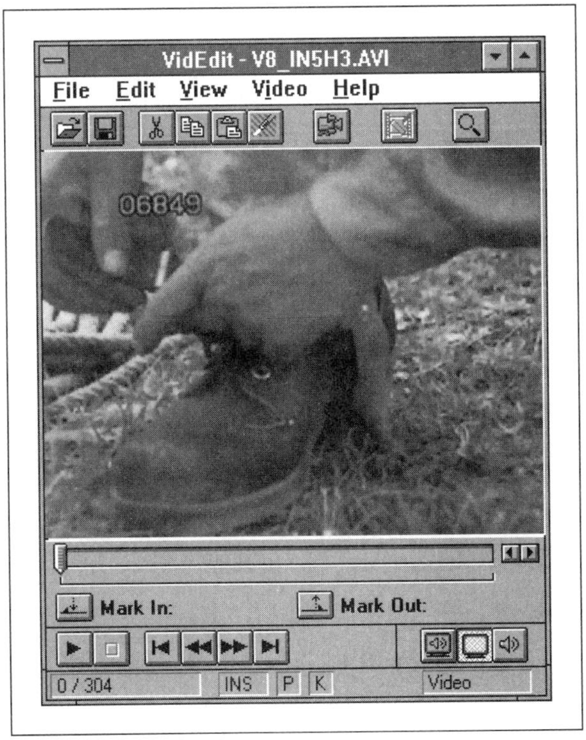

Figure 3.7 Video for Windows' video editor and compression manager, VidEdit

different resolutions, adjust video frame rate and color depth and change the audio parameters of the captured AVI file. You can also cut and paste, combine and synchronize audio and video files.

VidEdit is also the primary compression interface, which controls codec selection and compression parameters such as data rate, key frame interval, video quality and other settings. We'll refer to VidEdit frequently as we work through these compression parameters in subsequent sections.

Prior to the release of Video for Windows Version 1.1, Microsoft informally commented that they won't update VidCap and VidEdit, leaving the advancement of their capture and compression functions to third-party developers. Microsoft also released the new version as a developer's kit, rather than an end-user application. These signal Microsoft's intent to focus on the architectural aspects of Video for Windows, rather than the products. This new focus, coupled with the new APIs, makes supporting Video for Windows much more attractive to applications developers, both aesthetically and economically.

PalEdit, BitEdit and WaveEdit

Eight-bit images, videos and animations use color palettes to define which 256 of the over 16 million colors available on some computers are used in the image. This palette "instructs" the Windows system which colors to use during the display of these images.

PalEdit is a palette editor to edit, add or delete colors from a palette, or adjust the palette as a whole (Figure 3.8). This is useful when trying to lighten or sharpen the contrast of a video sequence, or otherwise change its appearance.

You can also use PalEdit to help construct an optimal common palette for groups of images, videos and animations that will share screen time during the course of a presentation. Without a common palette, the palette will shift or blink as the different videos play, or Windows will assign a common palette, typically with poorer results.

Figure 3.8 Video for Windows' palette editor, PalEdit

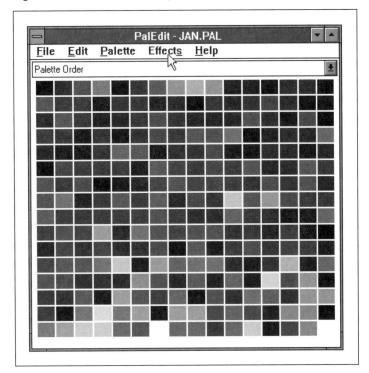

   The optimal color palette can be "pasted" into 8-bit video
files with VidEdit. BitEdit, another utility shipped with Video
for Windows, performs the same function for still images
(Figure 3.9). In addition, BitEdit performs color reduction,
where the color depth of 24-bit and 16-bit images are reduced
to 8-bit, so a common palette can be assigned.
   BitEdit handles these palette-related tasks quite adroitly and
will quickly become essential to those working in 8-bit environments. While BitEdit can also perform simple editing functions such as cut, paste, flip and rotate, serious image editing
should be left to full-blown programs such as Aldus
Photostyler and Adobe PhotoShop.
   WaveEdit is an application that edits WAV audio files
(Figure 3.10). Editing features include cutting and pasting,
fade-in and fade-out and amplitude adjustment. You can edit

Figure 3.9 Video for Windows' image editor, BitEdit

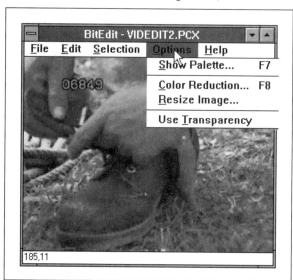

the audio frequency, sample size and number of channels to balance data rate against sound quality. WaveEdit also lets you record and digitize audio and store the captured audio as a WAV file.

Figure 3.10 Video for Windows' audio editor, WaveEdit

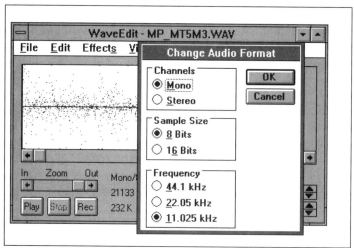

Unlike PalEdit and BitEdit, which perform a narrow range of critical functions extremely well, WaveEdit is a basic program useful only when other editing programs are unavailable. Typically, software shipped with most sound boards is more comprehensive and more feature-rich than WaveEdit.

CapScrn and VidTest

New in Video for Windows 1.1 is CapScrn, a screen capture utility (Fig 3.11). CapScrn can capture any screen image, or sequence of screen images, as well as audio.

CapScrn stores captured date as AVI files. Video is compressed using run length encoding (RLE) with interframe compression. Audio compression options include PCM, the standard, uncompressed waveform audio format; Microsoft ADPCM, audio compressed using the Microsoft ADPCM format; or IMA ADPCM, audio compressed using the IMA ADPCM format.

CapScrn opens up several interesting possibilities. For example, CapScrn can capture sequences of screen commands that demonstrate how to perform key application functions. Add audio and you've created a multimedia training video. You can also use CapScrn as a general purpose screen capture device that, through VidEdit, can deliver a bitmapped image in PCX, DIB, TGA, PICT or several other formats. We used CapScrn quite extensively to capture the screen images shown in this chapter and throughout the book.

Figure 3.11
CapScrn, Video for Windows' screen capture utility

Also new in Version 1.1 is VidTest, an application that tests a computer's video, audio and mass storage subsystems to determine how well the system will play video (Fig 3.12). VidTest works by playing an AVI video file and a WAV audio file and by streaming a file off the target storage device. It then measures the percentage of the computer's processing power used to perform these func-

Figure 3.12 VidTest, Video for Windows' system analyzer

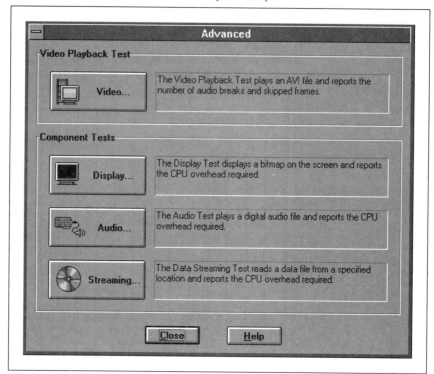

tions. VidTest then compares the computed values to standard figures to identify potential bottlenecks to video performance.

VidTest also plays video files and records the percentage of frames dropped and number of audio breaks. This is obviously useful when analyzing machine and codec performance.

Say Sayonara

In August 1994, Microsoft announced that they would EOL (End of Life) the Video for Windows application suite, meaning that they would discontinue the active sale of these tools.

Basic development of the Video for Windows architecture will, of course, continue.

While Microsoft did indicate that they may license the tool set to another company, nothing has been announced. Since these application specific tools are the simplest and most elegant ways to work with AVI files, you should track a set down for posterity. In the meantime, since most other Video for Windows applications emulate the functions of VidCap and VidEdit using the same basic controls, we'll illustrate these controls in the Video for Windows programs.

Video for Windows Summary

Video for Windows has one main challenger, QuickTime for Windows from Apple (Figure 3.13). In addition to its codecs, QuickTime is another multimedia environment that enables digital video playback inside Windows applications.

QuickTime dominates the Macintosh platform from both a codec and toolset perspective. However, while QuickTime is supported by many Windows-based video tools such as Premiere and MediaMerge, it really hasn't gained acceptance under Windows, in part because creating QuickTime movies, or MOV files, wasn't available under Windows. In June 1994, however, Intel released a utility that will create Indeo-based MOV files. It will be interesting to see how this changes the balance of power between QuickTime and Video for Windows.

Figure 3.13 Apple's Multimedia Environment, QuickTime

QuickTime movie player Version 1.1

® **© Apple Computer 1993**

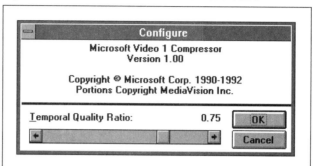

Figure 3.14 Video 1 from Microsoft and Media Vision

VIDEO FOR WINDOWS CODECS

Introduction

We've been bandying the names about for several chapters, now, so we should probably formally introduce you to the Video for Windows codecs. This discussion is fairly general. We'll get down to the nuts and bolts in Chapter 6, the Tour de Codec. If you're in a hurry to start compressing, you might want to jump directly to the Tour. If you're enjoying the flow and want to hear some early Video for Windows gossip, stick around.

In addition to the Video for Windows codecs, we'll also check out Scalable MPEG from Xing Technology Corporation. It's not only an extremely useful and accessible codec, it's an introduction to MPEG—a technology that will become increasingly important over the next 12 months.

To round out our discussion, we'll also briefly touch on two other codecs, Captain Crunch from Media Vision and Fractal Video from Iterated Systems.

Run Length Encoding

RLE was developed by Microsoft and included in the first release of Video for Windows. RLE uses run length encoding-based intraframe compression. This means that from a strict technical standpoint, RLE is the only lossless codec. We'll see if it lives up to its "lossless" title later on.

RLE is such a basic technology that Microsoft didn't even develop a logo for the product!

Video 1

Video 1 was developed by Media Vision and licensed to Microsoft for inclusion in Video for Windows release 1.0. Intraframe compression is based upon a proprietary lossy technology called MotiVE. Media Vision hasn't released any details about MotiVE, so we can't tell you much about it. Media Vision also hasn't explained the funky capitalization of the name. Presumably, the two are related.

As the first real Video for Windows codec, Video 1 gained a lot of momentum from early design ins. Generally, however, it has been surpassed in quality and pure compression by Indeo and Cinepak. Today it is only used in several specialized situations.

Indeo

Indeo was developed by Intel Corporation. Introduced in 1993, the codec was first incorporated into Video for Windows release 1.1.

Indeo uses vector quantization as its intraframe engine and probably has the most royal bloodline of all the codecs, being the progeny of DVI, or Digital Video Interactive. Intel purchased DVI from General Electric's Princeton-based Sarnoff Labs for $275 million in 1987.

Figure 3.15 Indeo from Intel

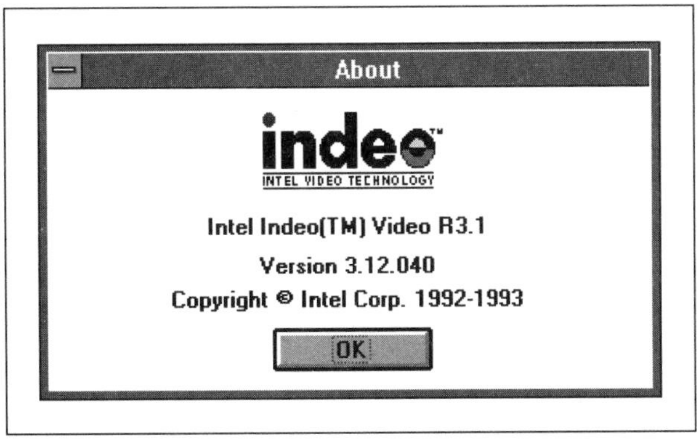

Since then, Intel has released three versions of Indeo. Version 2.1 was a strictly intraframe technology, largely focused on Intel's popular capture board, the Smart Video Recorder. In late 1993, Intel introduced Indeo version 3.1, which incorporated interframe compression as well. In May, 1994, Intel introduced beta samples of Version 3.2, scheduled for shipment in late summer 1994.

In addition to Windows, Indeo is available on the Macintosh and UNIX platforms. Indeo is also the technology underlying Intel's recent push into teleconferencing. As a result, Intel has a lot riding on Indeo and is backing the technology with significant marketing and development dollars.

Cinepak

Cinepak was developed by SuperMatch, a division of SuperMac Technologies. Cinepak was first introduced as a Macintosh codec and migrated to the Windows platform in 1993. Cinepak debuted in Video for Windows release 1.1.

Like Indeo, Cinepak uses vector quantization as its intraframe engine. Cinepak offers the widest cross-platform

Figure 3.16 Cinepak from SuperMac

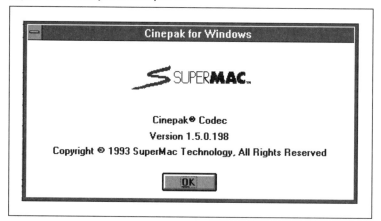

support of all the codecs. In addition to the Windows and Macintosh platforms, SuperMatch has ported Cinepak to the 3DO, Nintendo and Atari platforms. Because of its cross-platform support and outstanding performance, Cinepak is probably the most widely used codec on the market today.

QSIF MPEG

Scalable MPEG from Xing Technology Corporation was first introduced in 1992 and is the only software-only MPEG codec. Xing's Scalable MPEG uses MCI commands and is not Video for Windows–compatible. In addition, Xing doesn't interleave its audio and video streams, making it difficult to play synchronized video from a CD-ROM.

In the first release, Xing implemented what's called QSIF, or quarter-screen resolution, which translates to 160x120 screen size. During decompression, Xing interpolates the 160x120 video stream to 320x240, which causes a slightly fuzzy appearance. Xing also chose not to use delta frames, so the data stream consists solely of key frames, or "I" frames in MPEG terminology.

Figure 3.17 Scalable MPEG from Xing Technology Corporation

Xing's MPEG is the only technology that offers real-time capture to reasonable data rates. This niche is a small one, however, and Xing's products have not been widely commercialized. In August 1994, Xing released its full MPEG codec, reviewed in Chapter 15.

Fractal Video

Iterated Systems uses its own patented fractal technology as the interframe technology for its video products. While one of the first companies to show software-only video playback on DOS computers, Iterated has yet to significantly penetrate the computer market, with only a sprinkling of titles using fractal video.

Fractal video's primary claim to fame is scaleability. In theory, a relatively low resolution stream of fractal video can be zoomed to extremely high resolutions without significant quality loss. These claims garnered Iterated a $2,000,000 grant from the National Institute of Standards and Technologies (NIST) in 1992 to deliver a fractal decoder chip for high-definition television.

Iterated launched their first Video for Windows codec in January 1994. To compress, you have to purchase a developer's kit or contract for compression services at roughly $100 per minute of video. In 1992, Iterated partnered with TMM, Inc., of Thousand Oaks, California, for the resale of fractal video products. TMM offers both compression services and developer's kits. Unlike Iterated's Video for Windows technology, however, TMM's fractal video is based on the MCI command set. Both Iterated and TMM charge royalties.

Captain Crunch

Captain Crunch is a wavelet-based compression technology developed by rapidly falling star Media Vision. Wavelets have assumed the "technology darling" mantle from fractals, yet have failed to achieve significant commercial success for either still-image or video compression.

After Captain Crunch was announced in May of 1993 with extreme fanfare, mum has been the word. The product has never been officially released.

In July 1994, Media Vision entered Chapter 11 bankruptcy proceedings, another hi-tech Icarus falling from the sun. Even if Media Vision makes it through bankruptcy, it will probably focus on its core businesses and discontinue the marketing and/or development of Captain Crunch.

SUMMARY

1. Video for Windows was introduced by Microsoft to spawn multimedia development on the Windows platform. Judged by that standard, it has been a remarkable success.

2. Video for Windows is both an architecture and an application suite. The architecture centers around the AVI file format (Audio/Video Interleaved), which provides a common ground for capture card manufacturers, codec developers, video editing developers and other peripheral products.

3. There are two outbound architectures for AVI files, through Media Player and Object Linking and Embedding, and through the Media Control Interface (MCI) specification. Media Player is a simple interface that enables video linking to common applications like word processors and spreadsheets. The MCI structure, targeted more towards developers, is harder to use but much more powerful and flexible. An intermediate structure, presented by authoring developers who have implemented the Video for Windows MCI structure within their programs, provides a happy medium of flexibility and accessibility.

4. Video for Windows ships with four codecs, RLE from Microsoft, Video 1 from Media Vision and Microsoft, Cinepak from SuperMac and Indeo from Intel. Use of these codecs is royalty-free.

5. There are other Windows codecs that don't support the Video for Windows architecture.

6. The Video for Windows application suite includes seven applications as follows:

   (a) VidCap—video capture program;

   (b) VidEdit—video editing and compression manager;

   (c) PalEdit—palette editor;

   (d) WaveEdit—wave file editor;

   (e) BitEdit—bitmapped image editor;

   (f) CapScrn—screen capture program;

   (g) VidTest—system benchmark program.

   Microsoft stopped selling these tools in August, 1994. Anyone working with AVI files should track down a set, since in many ways they are the simplest and most elegant video tools available.

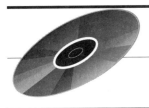

SOUND SYNCHRONIZATION

4

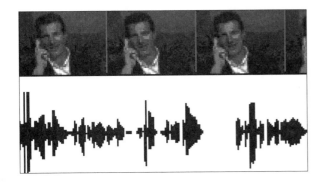

SOUND SYNCHRONIZATION

Sound synchronization, while not a compression function, is fundamental to the operation of codecs and video in general. For this reason, we'll discuss sound synchronization before delving into other compression and playback parameters.

Synchronization During Capture

Synchronization is the precise matching of the audio and visual elements of video, typically most important for talking-head sequences. It's also one of the "gotchas" in digital video, something you assume would occur naturally but seldom does. Unfortunately, while most viewers don't notice slight losses in video quality, out-of-synch video is conspicuous and often distracting. So learning how to achieve and maintain synchronization becomes a key video editing skill.

Let's review. During capture, an analog source feeds video frames into a frame grabber or video capture card, and the audio signals through a sound card. The capture software, either VidCap or another program, manages the process, ultimately creating an interleaved, and hopefully synchronized, audio/video file.

Figure 4.1 illustrates VidCap selecting capture at 15 frames per second. Figure 4.2 graphically represents one second of video captured at that rate, with the video frames on top and the digitized audio below. What happens if you capture the same sequence at 10 frames per second? This is shown in Figure 4.3.

If you compare the audio signal in Figures 4.2 and 4.3, you'll see that they're identical. The only change is that the sig-

Figure 4.1 VidCap set to capture 15 frames per second with audio

Figure 4.2 Video captured at 15 frames per second with audio

Figure 4.3 Video captured at 10 frames per second with audio

nal is spread over fewer frames, or that each video frame gets allocated a larger portion of the same sound signal.

Frame Dropping

What happens if your computer can't display 15 frames per second? The video loses audio synch, with the audio racing ahead; or, the audio "breaks" and stops, and starts waiting for the video to catch up; or, the codec "drops frames" and maintains synchronization while displaying less than the optimal number of frames per second.

You determine which occurs with the Media Player Skip Frames control shown in Figure 4.4, or the corresponding MCI control. This tells the codec to prioritize audio over video. When checked, all available processing power is expended to retrieve and play the audio, with any remaining power directed towards video decompression and display.

When you select "skip frames," video frames are retrieved and decompressed, but if audio playback is in jeopardy, video frames are "dropped." When a frame is dropped, the frame

Figure 4.4 Media Player selection to skip or "drop frames" if behind

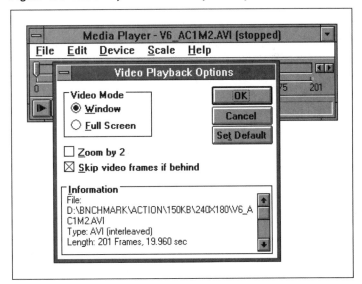

Figure 4.5 Frames 3, 8 and 11 dropped

immediately before the drop frame remains on screen until another frame displays. For example, in Figure 4.5, frames 3, 8 and 11 are dropped, and frames 2, 7 and 10 remain on-screen until frames 4, 9 and 12 replace them. This interrupts the natural flow of the video and causes a noticeable jerky appearance, especially during high-motion sequences.

Jerky or not, Video for Windows' bias towards audio is a great stride forward, because if you think frame dropping is unpalatable, you haven't heard an audio break. In most instances, our hearing is much more sensitive than our vision. For this reason, we're able to tolerate and even overlook some loss of video quality. However, virtually any loss in audio quality or audio break is immediately noticeable and annoying.

Back to the Media Player "drop frames" control. When you disable frame dropping, usually the audio breaks until the video catches up. Infrequently, the audio will race ahead and finish before the video. Either way, it's usually preferable to enable frame dropping to maintain some semblance of synchronization.

If you drop too many frames, your video won't look synchronized even if the audio and video are perfectly timed. We'll address how to balance these factors in Chapter 10.

POST-CAPTURE SYNCHRONIZATION

At times you'll need to add audio to existing video files. For example, when using video clip art, you may want to add sound to the selected video footage. VidEdit offers two options: you can insert a separate audio file using the

File/Insert command; or paste an audio file in the clipboard with Edit/Paste command.

Adding Audio—Insert Mode

VidEdit has two modes for inserting and deleting audio or video segments of an AVI file. In "insert" mode, audio or video segments added to the file will move the existing information forward to clear space for the new data. For example, in insert mode, if you add 300 frames of audio information to a 300-frame video, you'll end up with 600 frames.

When deleting audio or video information in insert mode, you delete the frames containing the information as well, assuming no other information was present in the frame. Information appearing after the deletion moves back to fill in the space.

For example, in a 300-frame file with interleaved audio and video, if you delete the first 150 frames of audio and video in insert mode, the first 150 frames would disappear. Frame one would be former frame 151, and your file would have only 150 frames.

If you delete only the audio portion of the first 150 frames, you'd still have 300 video frames, but the audio from frame 151 would move to frame one and the last 150 frames would have no audio.

Adding Audio—Overwrite Mode

Contrast this with "overwrite" mode. In overwrite mode, the audio or video segments added to the file *replace* the old information. If you add 300 frames of audio information to a 300-frame video, you'll end up with 300 frames, *even if there was audio in the original file*.

When deleting frames in overwrite mode, VidEdit removes the information in the frames, but leaves the empty frames. In our 300-frame file, if you delete the first 150 frames of audio and video in overwrite mode, you still have 300 frames, but the first 150 are empty. If you delete only the audio portion of the first 150 frames, the audio won't shift forward and your video will be a silent movie until frame 151.

Here are the two scenarios to watch out for. First, when adding audio to a file, keep your eye on the frame count. If it rises out of proportion to the audio inserted, you're in insert mode and should be in overwrite mode.

Second, when attempting to cut frames from a file you'll notice that the file doesn't get any shorter. You cut, and cut, and cut, and the frame count stays the same. This should clue you that you're in overwrite mode and need to switch to insert.

The toggle for overwrite mode is shown in Figure 4.6. To access this screen, select "Edit" and then "Preferences." Notice the mouse pointing to the OVR/INS preference indicator in VidEdit's main screen.

Figure 4.6 Overwrite toggle in VidEdit Preference Screen (Edit/Preferences). Note mouse pointing to OVR/INS indicator on VidEdit's main screen

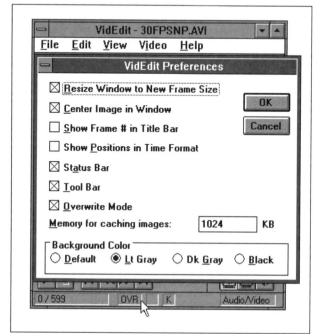

Synchronization

Once you've combined the two files, the VidEdit control shown in Figure 4.7 lets you synchronize the two signals by adjusting either the video speed or the video/audio offset. Use the audio offset control to move the audio stream forward (plus numbers) or backwards (negative numbers). Figure 4.7 shows the audio adjusted 100 milliseconds forward.

Our machines all have 8 MB of RAM, and we've found the Sample Playback of Video to be totally useless. With only 8 MB, your computer chugs and chugs and chugs and chugs, and then you get an Out of Memory error. Maybe you'll have better luck with 16 MB or higher.

With only 8 MB of RAM, the easiest way to test your synchronization is to save the file and play it again. Typically, if you attempt to play the video without first saving the file, video playback will be irregular with frequent audio breaks, and you won't be able to tell if you've helped or hurt synchronization.

Synchronizing audio and video is a trial-and-error process. Typically you get closer and closer until it's about right, but

Figure 4.7 Audio synchronization in VidEdit (Video/Synchronize)

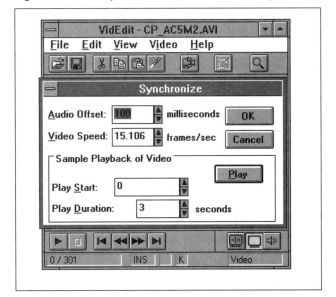

sometimes you seem to get further and further away until it's easier to trash the file. Use a temporary file until you're sure you have a keeper. When saving the file, be sure to use VidEdit's *no recompression* option.

No Recompression Option

Get familiar with the *no recompression* option in the Compression Method Selection box (Fig 4.8) very quickly—it will save you a lot of trouble. Here's why.

VidEdit doesn't have a compression button—you compress when you "Save" the file using the compression parameters loaded into VidEdit's Compression Options screen. When you load a previously compressed file, VidEdit also loads its original compression parameters. If you save the file again, VidEdit will *recompress the file to those parameters.*

If you load a file to add or synchronize audio, for example, and then save the file, VidEdit automatically starts to compress

Figure 4.8 The "No Recompression" option (Video/Compression Options)

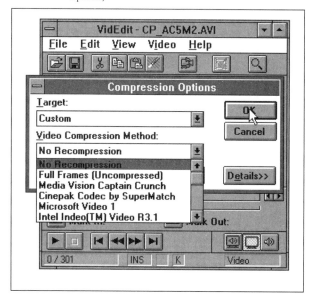

the file again. This double compression degrades video quality and destroys the original compressed file. You can avoid this by selecting the *no recompression* option.

It's pretty easy to spot your mistakes. Saving a previously compressed file using the *no recompression* option takes just a few seconds. Recompressing a file takes forever. So if you save a previously compressed file, and you notice it's taking a while to save, you've probably launched a recompression. Press Cancel to start over, and no harm is done.

Contrast this with *Full Frames (Uncompressed)*, another compression selection. This option takes the neatly compressed file that took hours to create and converts it back up into a raw, uncompressed file. This is another great way to blow away a file that took hours to compress.

Convert Frame Rate

Let's get back to the synchronization dialog box shown in Figure 4.7. In this dialog box, the Video Speed control affects the number of frames played per second. If you dropped the 15.106 frames per second shown in Figure 4.7 to 10, the video would contain the same number of frames but *would take 33% longer*. This wreaks havoc on audio/video synchronization, since the audio plays normally and the video plays 33% slower and finishes 33% later.

When you change Video Speed, you don't change the number of frames in the file, you change the speed that they play. Note that the file, CP_AC5M2.AVI, contains 300 frames, as shown in the bottom left-hand corner.

Contrast this with the Convert Frame Rate control shown in Figure 4.9. Here we've dropped the frame rate from 15 to 10 frames per second, a 33% drop. This tells VidEdit to eliminate every third frame. Look on the lower left-hand corner, and you'll see that the number of frames in the file did drop from 300 to 200. With this option, you don't change the speed of the video, you simply cut the number of frames displayed per second. Fundamental audio/video synchronization is the same,

Figure 4.9 The Convert Frame Rate Control

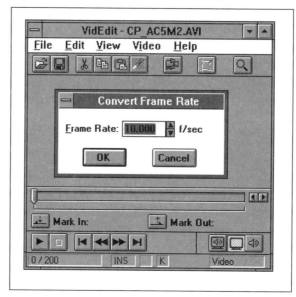

because the audio and video portions start and stop at the same time, even after adjustment.

Use the Change Video Speed option to speed up or slow down video, as in those instances where you want to create slow-motion video. Mind your sound synchronization, however, as it's easy to get out of whack. Use the Convert Frame Rate control to keep the same video speed, but play fewer frames per second, as when you want to adjust the video frame rate from 30 frames per second to 15.

Most of all, recognize that these controls can be confusing. It may pay to experiment on how these switches operate before clicking the wrong button on a production video.

SUMMARY

1. Sound synchronization is the precise matching of the audio and video elements of digital video, typically most important with "talking head" sequences. Sound synchronization is performed by the video capture software.

2. When computers can't decompress and display all the frames in the video stream, there are three options:

   (a) the video loses synchronization while the audio races ahead;

   (b) the audio "breaks" and stops and starts waiting for the video to catch up;

   (c) the codec "drops frames" and maintains synchronization while displaying less than the optimal number of frames per second.

3. Sound breaks are more obvious than drop frames. For this reason, most developers prefer to drop frames rather than risk audio breaks, prioritizing audio smoothness over video appearance.

4. Sound can be added after capture in VidEdit by Inserting or Pasting the file. The Overwrite and Insert modes produce different results that you should experiment with if you frequently must add audio to video files.

5. Normally, when you add audio to a video file, playback will be extremely jerky until you save the file and interleave the audio and video. When saving the file, be sure to use the No Recompression Option to prevent double compression.

6. The VidEdit Synchronize control lets you synchronize audio and video by delaying or advancing the start of the audio within the file. The Video Speed control within the Synchronize dialog box lets you increase or decrease the number of frames played each second, but doesn't change the total number of video frames. Use this control to create slow motion or accelerated video.

7. In contrast, the Convert Frame Rate control reduces the number of frames played per second by adding or discarding frames but doesn't change the base video rate. Use this control to reduce or increase the number of frames per second in the base video stream.

PLAYBACK
PLATFORM
CONSIDERATIONS

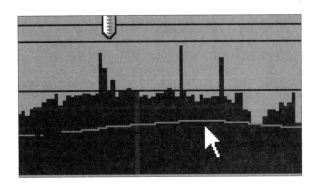

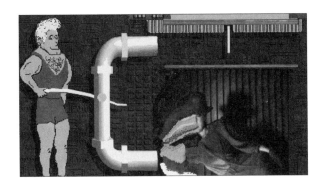

PLAYBACK PLATFORM CONSIDERATIONS

Before selecting critical video capture and compression parameters such as resolution and frame rate, you need to analyze how your *target system* will perform during video playback. The concept of target system is pretty simple—it's the playback system configuration towards which you are targeting the product containing the video. Sometimes your target system is every system in the universe—sometimes it's the sales jockey with the latest hot laptop. Either way, for your video to perform well on your target system, you have to identify your target and configure your video accordingly.

This section examines how various hardware configurations affect video playback. We'll look at processor and bus type, playback from CD-ROM and hard drive, and examine some new products and trends that will dramatically enhance video performance under Windows.

You'll learn how to buy and configure a computer for optimum video performance. You'll also see which codecs are taking advantage of these new technologies, and which aren't. Most important, however, is this—you'll finally answer the question, "Why does video always look better at trade shows than it does back in my computer?"

PROCESSOR/BUS

We've seen the load video places on the computer and how much work is involved in playing one video frame. So it's probably no surprise that video performance—measured in terms of display rate, or number of frames per second displayed—relates primarily to your host CPU. We'll post comprehensive playback statistics in Chapter 10 when we examine compression and playback parameters in detail. For now, we're happy to report that the results were surprisingly good, meaning that in some video modes, even 80386-based computers can play video at 15 frames per second at resolutions of up to 240x180.

Bus—Take the Local

The term "bus" refers to the circuitry that connects your computer's processor and main memory with components such as the hard drive and expansion slots. Since video requires extremely high-volume data transfers, bus type trails only the CPU in terms of importance to video performance.

The term "local bus" refers to a bus that's directly connected to the host CPU and runs at the same bandwidth and clock speed. Originally, all processors and buses were local and ran at the same speed. Starting with the IBM AT in 1984, computer architectures changed to allow the processor to run faster than the outlying circuitry. This was the birth of the "expansion bus" embodied in what came to be called the Industry Standard Architecture, or ISA bus. Essentially the ISA bus freed the host processor to run faster than the bus, which was great for the processor but obviously not so great for the bus.

Let's expand on a couple of terms. As we've discussed, bandwidth describes a system's ability to manage data flow. From a bus and processor perspective, it's useful to think of the bus as a highway. An 8-bit bus has one lane, a 16-bit bus two lanes, and so on. From a CPU perspective, the more lanes you have, the more data you can crunch at one time. Similarly, from a bus perspective, the more lanes, the more data you can carry at one time (see Fig 5.1).

Bus speed, usually expressed in terms of megahertz, or MHz, is like the speed limit. Theoretically, an 8 MHz bus transfers data at about 8 MB/s per 8-bit channel. In practice, actual bus performance relates to architectural factors, so you can't precisely calculate maximum bus performance based solely on speed and bandwidth.

By and large, however, a 32-bit bus will perform twice as fast as a 16-bit bus, and a 33 MHz system will operate four times quicker than an 8 MHz system. Table 5.1 shows the speed and bandwidths of Intel architected processors. It's easy to see how opening up the bus to advanced processor architectures will dramatically increase performance.

However, as processors got faster, the expansion bus was mired at 16-bits bandwidth and a maximum clock speed of 8 MHz. Architectural bottlenecks limited the ISA bus to a peak throughput of about 8 MB/second and an average sustained throughput of about 2.5 MB/second. So whether you had a 16-bit 80286/10 MHz or a 32-bit 80486/33 MHz, if the bus was ISA, the speed at which data traveled through the bus was the same.

Another limitation of the ISA architecture is that it forces the host CPU to manage all data transfers. This accounts for the hourglass you're so accustomed to seeing when saving files to disk or retrieving them—when storing or retrieving, the CPU is dedicated solely to that task. To make matters worse, during data transfer, the CPU slows to the speed of the expansion bus, converting your turbocharged 80486/66 to a putt-putt 80286/8.

IBM took the first shot at turbocharging the bus with the proprietary and backwards-incompatible Microchannel archi-

Table 5.1 Processor bandwidth and speed

Processor	Bandwidth	Speed
80286	16-bits	up to 10 MHz
80386	32-bits	33 MHz
80486	32-bits	100 MHz
Pentium	64-bits	100 MHz +

tecture (MCA) launched with the PS/2 line of computers in 1987. The MCA architecture featured a 32-bit bus that ran at 10 MHz, which more than doubled theoretical performance. More importantly, however, was that the new architecture allowed processors other than the CPU to control the bus, enabling it to operate independently of the speed and bandwidth of the host CPU. The concept of "bus mastering," where devices other than the host CPU control the bus, was born.

However, the price of this added performance was incompatibility with all previous bus standards. Back in 1987, demands on the bus were dictated more by programs such as Lotus 1,2,3 and WordPerfect than by video, and the ISA bus architecture was sufficient for most users. For this reason, the MCA architecture was a technological step forward embodied in what turned out to be a huge marketing step backward.

Compaq and others responded with the EISA (Expanded ISA) bus. This backwards compatible standard expanded bus bandwidth to 32-bits but kept the maximum speed at 8 MHz. EISA could manage burst transfers of about 33 MB/second, but average throughput hovered at around 10 MB/second. More importantly, as a platform, EISA expanded on the concept of bus-mastering, spawning the introduction of many bus mastering devices.

Once again, for most users, the ISA bus was just fine. EISA became synonymous with server and never sold in quantities sufficient to reduce prices to commercial desktop levels. Besides, back in the late 1980s, color depth hovered at about 4-bit and the average full-screen bitmapped image was under 50 kB in size. Then Macintosh-envy took hold and everybody started walking around saying "Gooey" (GUI). Diamond Computer Systems build a company around the Diamond SpeedStar, a 24-bit card video graphics card that cost under $200. Microsoft announced Windows 3.0 and then 3.1 with unprecedented fanfare. All of a sudden, while no one was looking, graphics were everywhere and the average full-screen bitmapped image was 750 kB in size.

Then digital video arrived.

VESA Local Bus (VLB)

The first modern local bus was the VESA local bus, a standard promulgated by the Video Electronics Standards Association. This standard was intended to *add* local bus capabilities to the standard ISA, rather than replace the bus.

The optimal and most common VESA configuration is three VESA slots,a bandwidth of 32-bits and transfer speed of 33 MHz. This translates to a maximum transfer capacity of 132 MB/second and average sustained throughput of about 17 MB/second.

Figure 5.1 graphically depicts the difference between ISA and VESA local bus. Not only does the local bus have more lanes, it also has a faster speed limit. This translates to faster data throughput, which really helps during during the second bus transfer from main memory to the video graphics card.

 Animation—Double-click on BUSWARS.AVI in the chap_5 subdirectory to see this animation.

That's because after the video decompresses in main memory, it expands back to raw, uncompressed data rates. A 320x240x24-bit video playing at 15 frames per second enters main memory at 150 kB/s and leaves at 3.456 MB/second, which exceeds the sustainable throughput of the ISA bus. This stops ISA-bus computers in their tracks.

So, it's clear. If you're buying a new computer, the local bus is definitely the magic bus. The question is which local bus design, VESA or PCI?

PCI Bus

Peripheral Component Interconnect, or PCI, is a local bus architecture developed by Intel to compete with the VESA local bus standard. Like the VESA bus, PCI offers maximum transfer rates in the 132 MB/second range. PCI also offers several key additional benefits, including the promise of plug-and-

Figure 5.1 ISA bus with a 16-bit bandwidth and 8 MHz speed limit compared to the 32-bit VESA local bus with 33 MHz

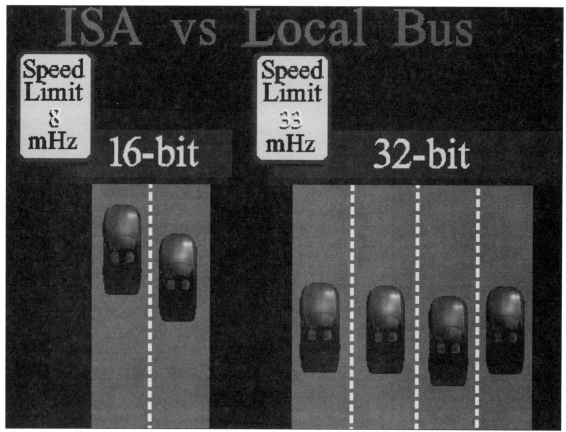

play peripheral installation. Anyone who knows an interrupt from an I/O port is a big fan of plug-and-play, which essentially means that all peripheral cards are self-configuring—no DIP switches or jumpers. From a video perspective, however, the key benefit is bus co-processing.

Figure 5.2 introduces "Buster," our animated rendition of the bit-pumping bus-mastering PCI chip from Intel. Overall, bus mastering makes PCI more efficient than VESA-equipped computers. Even if the theoretical maximum speed is the same, the bus coprocessor does the grunge file transfer work, not the host CPU, which is free to perform other tasks, such as decompressing video at a faster rate.

Most current PCI implementations have a 32-bit bandwidth, but 64-bit versions leveraging the Pentium's 64-bit bandwidth will soon be available. The theoretical maximum throughput of the 64-bit PCI bus is 264 MB/second. On the other hand, the VESA committee is pushing their 2.0 specification, which boasts a theoretical maximum of 400 MB/second.

Video—You can view a live demonstration of this process by double-clicking on BUSCOPR.AVI in the chap_5 subdirectory on the CD-ROM.

Who'll win in the great local bus wars—VESA or PCI? I'd have to guess Intel and PCI. The computer industry's technical cognoscenti seem to consider PCI technically superior, and plug-and-play is starting to become a key industry buzzword. Unlike the VESA standard, PCI is designed for use with multiple processors and is processor-independent to the point that it will work with non-Intel platforms like Digital Equipment Corp's Alpha chip.

More important is the "who cares the most" analysis. The VESA committee members don't really care which standard wins—they'll happily sell PCI or VESA cards. On the other

Figure 5.2 Buster—the PCI bus-mastering chip that offloads bus transfer functions from the host processor

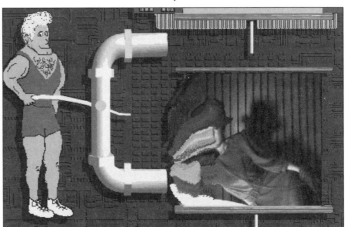

hand, when Intel gets their development, marketing and corporate ego behind a standard, they usually prevail. Intel reports that more than 200 companies have adopted the PCI specification, so they're winning a lot of design-ins. So when it's time to buy a Pentium, I'd feel pretty comfortable making it PCI.

ISA Strategies

Back to reality for a moment. The world isn't solely composed of PCI Pentiums, or even 486 VESA local bus machines. There's a ton of ISA bus computers and even the rare 80386 out there. How do we work best with these computers?

Once again, we'll post complete statistics in Chapter 7 when we address how to select resolution and frame rate for your compressed video files. For now, we'll focus on how display resolution affects video playback.

As you probably know, most video graphics cards display at multiple resolutions and color depths. Back in 1992, inexpensive 24-bit graphics cards started to appear, led by the aforementioned Diamond Speedstar. Without question, these cards improve the appearance of video and still-image graphics alike. However, they can slow video's display rate to an absolute crawl. Table 5.2 shows why.

When you take advantage of the enhanced color provided by 24-bit cards, you create additional data, which takes space. Playing video on a 24-bit system requires three times the data

Table 5.2 Display rates at different color depths

	8-bit display	24-bit display
Uncompressed frame size (320x240)	77 kB/s	230 kB/s
Data rate per second (320x240)	1.15 MB/s	3.5 MB/s
Display rate (80486/33 MHz ISA-bus computer playing 320x240 Cinepak file.	15 fps	1 fps (yes, 1)
Display rate (80386/33 MHz ISA-bus computer playing 240x180 Cinepak file)	15 fps	1 fps (ditto)

necessary for playback on an 8-bit system. This often bounces the video data rate up against the capacity of the ISA bus, which, as shown in Table 5.2, cripples your display rate.

The simple answer? If you distribute your video, advise users with older and or slower computers to play the video in 8-bit mode. This detracts from the quality of the individual frames, but maximizes video display rate, which is usually more important. It also opens up the Pandora's Box of palette issues addressed in Chapters 12 and 13.

CD-ROM vs. Hard Drive

Processor Overhead—Display Rate

Don't expect video to run as fast from a CD-ROM as from a hard drive (Table 5.3). Even if the data rate is under CD-ROM bandwidth, retrieving data from CD-ROMs takes more processing power than retrieving from hard drives. This means less power for decompression and display, which translates to fewer frames per second and jerkier video.

This relates to the transfer rate of CD-ROMs—usually 150–300 kB/s, as compared to 2.5 MB/s for the slowest hard drives and to the fact that the host CPU actually performs the transfer. For example, when playing 150 kB/S video on a single-spin CD-ROM, it takes one full second to retrieve the video for each second of video. In other words, the video transfer is ongoing for the duration of the clip—which takes a lot of CPU cycles. The same transfer takes under 40 milliseconds for each second of video on a slow hard drive, about 1/16 of the time. Obviously, the burst transfer from the hard drive is much easier on the host CPU.

Table 5.3 Processor overhead required to retrieve files. Tests performed on 80486/66 MHz VESA local bus computer with 8 MB RAM

	Hard Drive	CD-ROM
150 KB/S transfer	11.1%	21.10%
300 KB/S transfer	42.10%	59.6%

When your computer is already struggling to decompress and display video, the additional overhead imposed by the CD-ROM transfer translates directly into fewer frames per second.

Results will vary for different CD-ROM and hard drives, but CD-ROMs almost always require more processor overhead than hard drives. For example, the MPC-1 specification promulgated by the Multimedia PC Council requires that the CD-ROM draw no more than 40% of the processor overhead at 150 kB/S transfer rate. Double-speed CD-ROM drives under the MPC-2 specification can draw no more than 60%.

PC Magazine recently tested CD-ROM drives and found a surprising discrepancy in performance (see February 22, 1994). On similar files, for example, single-spin drive scores ranged from 29% to 100% of processor utilization. Double-spin drive scores ranged from 13% to 90%. Incidentally, the CD-ROM drives that consistently performed the worst from a processor overhead standpoint were parallel port connected drivers. While they are the height of convenience, these are not the drives with which to demonstrate your latest video creation.

This means two things. First, when you purchase a CD-ROM drive, you should check the percentage of overhead required to retrieve a file. It should at least be under the MPC-specified rates, and hopefully much better. Second, when you're selecting compression parameters for video to be played from a CD-ROM drive, remember that it won't play as well from the CD-ROM as it does from your hard drive, even if the file is under 150 kB/second.

Bandwidth Limits

CD-ROMs have fixed bandwidth capabilities. Single-spin, or 1x, drives transfer at a maximum of 150 kB/second. Double-spin, or 2x, drives transfer at about double that rate, while QuadSpin drives transfer at about 600 kB/second. When you develop for CD-ROM distribution, your target system definition becomes critical, because your decision can absolutely exclude certain users from viewing your video.

For example, if you assume your target system is a PCI Pentium, 386 owners will likely still be able to play your videos. Frames may drop, it may look like our proverbial slide show, but you can watch it. However, if you put video files with a 300 kB/second bandwidth on your CD-ROM, single-spin owners can't play the video without audio breaks.

We recommend developing for the single-spin platform, meaning that you compress your files to under 150 kB/second. The question is, how closely must you adhere to the fixed 150 kB/second limit?

Here's how we look at it. Although most CD-ROM drives buffer to some degree, as does MSCEX.EXE, your Windows CD-ROM driver, it's best to consider the CD-ROM drive a totally fixed bandwidth pipe. You should definitely make certain that no video segments exceed the bus bandwidth—otherwise, you won't be able to retrieve the data fast enough to play the video. In the best case, you'll end up dropping a bunch of frames. In the worst case, your audio will break up.

How do you make certain that all the segments in the video file are under the single-spin CD-ROM rate? The codecs don't compress every frame uniformly through the file—there's typically data spikes throughout the file.

Well, the following is a paid political advertisement for a product you already own—VCS Play, on the CD-ROM. It'll be more interesting if you follow along. I'll wait till you boot your PC.

No-no-no, take your time, I'll wait. Ready? Here goes.

Several Windows programs purport to tell you the data rate of your video clip. Let's start with File Manager. That'll give us a gross measurement of the video clips bandwidth. We'll be looking at file CP_AC3M3.AVI. You can find it in the chap_5 subdirectory of the disk. First we'll check out the file size in File Manager, usually found in your Main Group. Double-click on File Manager, click on the CD-ROM drive, and then on the chap_5 subdirectory, and touch the file. Figure 5.3 shows the file.

If we divide file size by duration, File Manager tells us that this file has a bandwidth of 150,814 bytes/second (3,016,280 bytes/20 seconds). Looks like we'll have to throw it back. Too bad, because it took about an hour and a half to compress.

Figure 5.3 Since cp_ac3m3.avi is 20 seconds long, File Manager tells us that this file has a bandwidth of 150,814 bytes per second

Maybe we should get a second opinion. Let's check out Media Player.

Media Player usually hangs out either in your Video for Windows Group, Accessories Group or both. Double-click and let's load the file.

Figure 5.4 shows 145 kB/second—must be livin' right! Media Player will typically show a lesser value than that derived by dividing file size by file duration. Video files aren't all audio and video; there are also overhead components that never leave the disk during playback.

Media Player measures only the retrieved information, which makes sense, because if you don't have to retrieve it, it doesn't take any bandwidth.

Figure 5.4 Media Player shows a bandwidth of 145 kB second

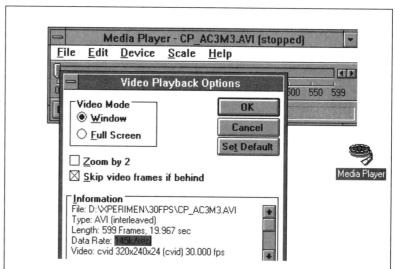

Figure 5.5 VidEdit shows 274 KB/s

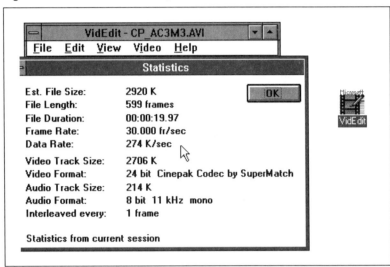

Maybe we won't have to stay till midnight, after all. But just to be certain, maybe we ought to check just one more place. Let's try VidEdit. You won't have this program unless you own Video for Windows. But if you have it, let's see what it says.

Figure 5.5 shows the bad news—274 kB/second. Better call the hubby and tell him not to wait up. But hold on. What about that . . . that . . . that VCS program we bought last week. Doesn't that have some kind of cool feature that *shows* us exactly what the data rate is? Where is that CD-ROM?

Three minutes and one seamless installation later. Double-click on the icon in the VCS subdirectory.

Voila! Or should I say, "Whoomp, there it is?"

Figure 5.6 shows VCS Play's frame profile feature. This is a histogram showing the size of each frame in the file. The wavy gray line tracks the per second data rate and plots it against the black 150 kB/second line that starts on the right-hand side.

Uh, oh! Looks like the mouse cursor is pointing towards a data spike that will prevent smooth playback from a CD-ROM.

And, if you play the file from a single-spin CD-ROM drive, you'll see that it hangs like a bad joke at a cocktail party.

OK, advertisement's over. Pretty cool feature, though, wouldn't you say? We had more than 250 files to put on the

Figure 5.6 A straight answer at last, thanks to VCS Play's
Frame profile

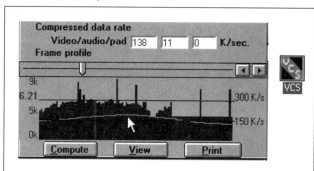

Video Compression Sampler CD-ROM, and we couldn't get a straight answer from the Video for Windows' tools either. So we built the frame profile into VCS Play and checked every file. So should you, because if you don't, there may be data spikes that prevent smooth playback on your target CD-ROM drives.

DISPLAY CONTROL INTERFACE—DCI

One of the major problems with video in Windows is Windows itself. As we saw in Chapter 2, Windows, like all graphical user interfaces, imposes significant burdens on the host CPU. That's why most DOS applications still run much faster than their Windows counterparts.

Once again, video imposes its own tremendous demands on the computer. Combine that with Windows overhead, and you've got a real bar against high performance. Unless, of course, you can find some way around Windows.

One way around Windows is Death, a special Video for Windows call that kills Windows Graphics Device Interface. Also called Full Screen mode, this crude mode lets the codec avoid GDI and write directly to video hardware, with a real boost in frame rate. Let's experiment and see just how much.

Figure 5.7 Selecting "Full screen" playback in VCS Play file selection screen

Playback parameters
Scale   Video mode   Other options   Display side
- .5x   ⊙ 1x   ○ Windowed   ☒ Skip frames   ◄ ►
- 2x   ○ 4x   ⊙ Full screen   □ Don't buffer

Set Playback Default

Load VCS Play. On the left-hand side load In_AC3m3.AVI in the chap_5 subdirectory. This is a 30 frame per second Indeo file that should stress out your playback system. On the right-hand side, load the same file, but select "Full screen" mode in the compression parameters on the bottom of the file selection screen shown in Figure 5.7.

Play the two files, first on the left and then on the right. Once you play the file on the right, you'll immediately see why they call this mode "Death." Death is a very crude video mode that converts your system into VGA Mode 13. GDI goes away, and the video plays in 320x240 very coarse pixels at 8-bits color depth. It's not a bad look, especially from far away, but you lose all the nice buttons and other Windows attributes. But, it's a real quick and dirty way to see how much GDI overhead actually costs in terms of frames per second.

Table 5.4 shows what I got on my trusty Gateway 80486/66 MHz VESA local bus computer, playing back from a single-spin CD-ROM drive.

Pretty dramatic difference! Almost 300% faster video when you avoid GDI and talk directly to video hardware. However, the disadvantages of this approach are fairly obvious—you lose the Windows interface.

Table 5.4 Display rate of 30 fps Indeo file in Window and Full Screen death mode

	Left Side (Window mode)	Right Side (Death mode)
Display rate	10 fps	28 fps

There are techniques allowing codec developers to write directly to video hardware and avoid GDI while preserving all the Windows attributes. Several codec developers have adopted these techniques and have speeded up decompression—in some cases. However, these techniques are not standardized and won't work with many graphics cards, a real problem for developers and end users. A standard is coalescing. To understand where it's going and why, let's take a walk down memory lane.

Video Driver Overview

In the beginning, there was DOS, and it was good. All DOS applications that displayed bitmapped graphics used their own proprietary format. Display was simple because video modes such as EGA and VGA were standard among the video cards. Applications developers had to write one driver for each video card, and that was it.

As video card developers expanded their capabilities, they ventured into nonstandard video modes, such as SVGA, XGA and others. Applications developers seeking to use these nonstandard modes had to write a driver for each card—a daunting task when you consider the sheer number of video cards and the frequency of updates.

As we've seen, Windows addresses this problem with the Graphics Device Interface, or GDI, to serve as the exchange medium between for applications and peripherals, including video cards. In this role, GDI accepts device-independent

Figure 5.8 Windows Graphics Device Interface

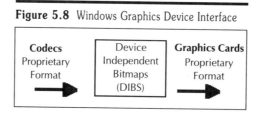

bitmaps, or DIBs, from applications and converts the image to the proper video card format with drivers supplied by the video card manufacturers (see Fig. 5.8).

Applications developers work in their proprietary format and convert to DIBs to render images to the screen. No video card–specific drivers are required. Ditto for video card manufacturers, who write one Windows driver that takes in DIBs and outputs graphics in their proprietary format. This lessens the burden on all developers and provides some assurance to end users that their applications and peripherals will work together.

The price we pay for this compatibility is speed. As we've seen, GDI costs can drop your frame rate by as much as 66%.

As we mentioned, several codec developers wrote drivers allowing them to avoid GDI and talk directly to the video card. However, as in the old DOS days, they have to write a driver for each supported graphics card, and have to update the drivers each time the graphics card manufacturers update *their* drivers. Unsupported graphics cards would operate through GDI with the concomitant drop in display rate.

This imposes a significant development burden on the codec developers, who were always one or two video boards and two to three drivers behind the installed base. While multimedia developers enjoyed the speed increase, they were turned off by the administrative hassles of driver installation and maintenance. After all, they were the ones the customers would call to get the updated driver, not the codec developer. Finally, end users rushing to buy the latest graphics board could find their video running slower on the new board if their old board was supported and the new one wasn't.

Predictably, most developers shied away from these nonstandard approaches in droves, and users tolerated suboptimal performance under Windows. Then, in August 1993, sensing a feature/function vacuum that translated to the opportunity for another

STANDARD

and another acronym to boot, Intel jumped into the fray with the Video Device Interface, or VDI.

Figure 5.9 VDI functions

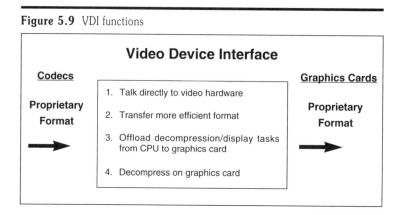

At its lowest level, VDI defines a protocol for codecs and video cards to work around GDI (Fig. 5.9). It's the equivalent of the old EGA/VGA drivers—by supporting VDI, a codec developer can write directly to the frame buffer of all VDI-compliant video boards, and avoid GDI. Video card developers support one specification and immediately can add "Video for Windows accelerator" to all spec sheets, ads and brochures.

At higher levels, VDI off-loads functions normally performed by the host CPU to the video board processor—from Mikey to Vinnie—which makes the overall decompression/display task more efficient.

Operationally, the codec polls the video graphics board upon installation, essentially asking, "What advanced functions can you perform, big boy?" The board answers, "Well, if you're real nice, you can talk directly to my video frame buffer," and together they go off, into the sunset, happily avoiding GDI. Everybody wins.

Intel announced VDI in August, 1993. However, Microsoft, concerned about possible architectural conflicts with Chicago (now Windows 95), asked Intel to work towards a joint specification.

By fall COMDEX, Intel and Microsoft released a joint standard called DCI, for Display Control Interface. DCI is a certifiably good standard. It will boost performance of all software-only codecs and doesn't require any additional hardware. Most major graphics card vendors have pledged support and are announcing—but not shipping—products as I write this chapter. For this reason, we haven't tested any DCI-compliant boards.

However, when Intel announced VDI, they claimed that VDI boosted performance by 100%, a 2x performance boost for VDI systems. Our tests showed a 300% increase from simply avoiding GDI, so Intel's claims are believable. In any event, DCI support belongs on your checklist of 'must-have' features for graphics boards.

From a codec perspective, Intel's Indeo will obviously lead the DCI support charge, and all other codec developers will probably follow suit. The net result is a real "win-win," better performance for all at no extra cost.

VIDEO CO-PROCESSORS

Video co-processors are chips on graphics boards that accelerate video performance by completing tasks that otherwise would have to be performed by the host CPU. This helps in two ways.

First, it reduces the work performed by the CPU, freeing it for other functions. Second, by decompressing and converting on the video card, video co-processors decrease the amount of data transferred through the bus, further cutting the CPU's workload.

Video co-processors are the wave of the future. Working within the DCI framework, they help deliver more frames per second at larger video resolutions. Let's take a closer look.

Decompression Detail

Let's break the decompression task into a little bit more detail. This will help us understand how video co-processors boost video performance.

As you know, we typically start with a compressed stream of about 150 kB/second. This gets decompressed to YUV format, which is a color-reduced but "video-friendly" format. This enlarges the video to about 1,152 kB/s. The YUV stream then

Figure 5.10 Decompression steps and corresponding bandwidths

Decompression Detail	
Task	**Data Rate**
1. Start with compressed video	150 kB/S
2. Decompress to YUV	1,152 kB/S
3. Convert to RGB	3,456 kB/S
4. Scale to full screen	13,824 kB/S

gets converted to a device-independent bitmap for transfer to the Windows buffer. This RBG format weighs in at about 3,456 kB/s. Finally, let's assume we scaled the video to full screen. This would quadruple the data to a crippling 13,824 kB/s, tough for even the fastest bus to maintain.

Smooth Scaling

On our benchmark 486/66 MHz VESA bus computer, you'd be lucky to yield 1 frame per second. Slip in a VideoLogic Movie 928, a video graphics card equipped with VideoLogic's PowerPlay 32 chip, the first commercially available video co-processor, and you'd achieve the full 15 frames per second. With the properly configured file, you could even get 30 frames a second at full screen. Here's why.

When operating without a video co-processor, the CPU performs all the conversion and scaling functions. It also has to transfer almost 14 MB/second to the video card—which really slows the process. Finally, when Windows scales to full screen, it simply replicates every pixel four times, which creates the blocky effect shown on the left-hand side of Figure 5.11. The result? Slow, ugly video.

With a video co-processor, the host CPU sends the 320x240 stream out to the video card, reducing the bus transfer from just under 14 MB/second to a more digestible 3.5 MB/second.

The video co-processor scales the video to full screen. Because such chips are application-specific by design, they scale very efficiently, and also filter and interpolate the data, rather than simply replicating pixels. This causes some fuzziness, which has the beneficial effect of masking small compression artifacts. Overall, video co-processing produces faster, higher-quality video.

Interestingly, as tested, the Movie 928 didn't accelerate video played back at its native resolution, meaning that a 320x240 video file played back *at 320x240* wasn't accelerated. In fact, at the video's native resolution, the Movie 928 performed about the same as other graphics cards without video coprocessors.

Figure 5.11 Compare pixel replication (left side) with interpolative scaling produced by the VideoLogic Movie 928 and PowerPlay 32 chipset

The reason? Because the DCI specification was not available when the card was completed, the card *didn't* avoid GDI, and *didn't* perform the YUV to RGB conversion. The only function the Movie 928 accelerated was zooming.

However, once DCI is available, most video cards will allow the system to avoid GDI, convert YUV to RGB *and* scale the video. The host CPU will only have to retrieve and partially decompress the data. This will turn all computers into high-performance video stations.

Figure 5.12 illustrates how DCI and video co-processors take work away from the host CPU. Without DCI, the CPU decompresses, converts to DIBs and scales, then sends a hefty 14 MB/second through the bus to the video card. With DCI, the processor decompresses and then hands off a svelte 1.15 MB/second to the video card for color space conversion and scaling.

Once again, the codec polls the video card to "negotiate" their respective tasks. Obviously, if either the codec or coprocessor chip doesn't support DCI, this negotiation is impossible. In these cases, the processor perform all decom-

---

Figure 5.12 Work transferred from CPU to video card with DCI and Video co-processors

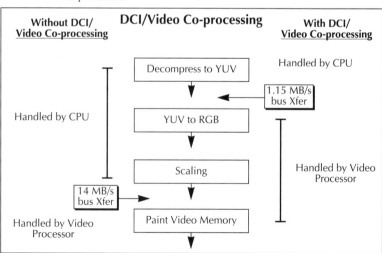

pression and display functions with the attendant drop in frame rate.

All in all, however, look for video co-processors to be as pervasive as DCI support. In fact, it's probably a safe bet that virtually all graphics cards will feature some form of video co-processing by the middle of 1995, which bodes well for all video users.

One interesting question is whether video co-processor performance will ultimately be codec-specific. Obviously this doesn't matter for avoiding GDI or simply scaling DIBS to larger resolutions. However, as graphics cards/chips move higher along the DCI functionality ladder and start to assume actual decompression functions, codec-specific instructions in the video hardware could really enhance performance.

As you read these words, codec vendors are spending serious marketing dollars convincing video chip developers and video card manufacturers to embed their algorithms in their video chips or video cards. Remember the '60s movie *Bob and Ted and Carol and Alice?* That's a good description of the video market right now. For example:

- Creative Labs, who licensed Cinepak from SuperMac and sells all of SuperMac's Windows-based products, recently announced two Indeo-based capture boards with embedded I-750 chips. So did ATI.
- Weitek, recently announced a chip that supports Indeo, Cinepak, Captain Crunch, MPEG, and Motion-JPEG.
- To complete the circle, SuperMac, in addition to Weitek and Creative Labs, has Cinepak licensing deals with Apple, Cirrus Logic, Sega, 3DO and Atari.

By mid-1995, all major video card and video chip developers will have announced products that co-process Indeo, Cinepak and maybe MPEG. These devices will boost performance of the supported codecs far beyond that offered to unsupported codecs. While a boon for end users, this level of functionality is also a virtual bar to smaller codec vendors who don't have the marketing muscle to convince graphics developers to embed their algorithms.

Thus, video co-processors, which along with GDI will boost desktop video into the mainstream, also sound the death knell for smaller codec developers, who in large part were responsible for launching the market in the first place.

SUMMARY

1. Advanced processor and bus architectures acellerate video playback dramatically. Local bus is the Magic Bus where video is concerned. Select PCI over VESA in the upcoming bus wars.

2. Those with ISA systems should play video in 8-bit mode to maximize performance.

3. Video playback from CD-ROM generally is slower than playback from a hard drive. When developing video for CD-ROM playback, ensure that the entire file is under CD-ROM data rates.

4. DCI and video co-processors will dramatically enhance video performance over the next several years. Any graphics boards purchased from now on should support DCI and should have either a video coprocessing chip or an upgrade socket.

TOUR DE CODEC

6

TOUR DE CODEC

Each time you open the VidEdit Compression Options Screen, four or five codecs stare you in the face, begging for selection. "Use me, use me," you can almost hear them cry. Happily, it turns out that all Video for Windows codecs have their strengths and niches. While you'll almost certainly end up using Cinepak and Indeo for most tasks, Video 1 and RLE are critical weapons in your fight for video perfection.

Along with the Video for Windows codecs, we'll also review Scalable MPEG from Xing Technologies, another software-only codec with application-specific strengths. Finally, we'll take a quick look at Captain Crunch, the promising, Wavelet-based codec from MediaVision that, unfortunately, never got shelf-space.

Finally, we'll look at the legal end. Once you select a codec, how do you get your hands on it, and what's it gonna cost you?

THE GREAT CODEC COMPARISON TEST

Comparison Criteria

Intuitively, the measure of any codec is the ability to faithfully recreate the appearance of the original video. There are two obvious components—the quality of the individual frames and the display rate. All of the benchmark videos are contained in the chap_6 subdirectory on the accompanying CD-ROM—you can follow along and judge quality and display rate yourself, or rely on what we show here. We'll post the display rates we achieved on our benchmark Gateway 486/66 VLB bus with 8 MB of RAM with an ATI Ultra Pro graphics board. All clips were played from a single-speed CD-ROM in 24-bit graphics mode.

We'll analyze four types of video files, a talking-head sequence and a high-motion sequence, both at 320x240 resolution; a 240x180 talking head shot; and an animated sequence converted to AVI format.

Let's review before getting started. Most codecs use both interframe and intraframe compression. As you recall, interframe compression is compression achieved by eliminating interframe redundancy, or image segments that remain static between frames. Intraframe compression is compression achieved solely within an individual frame.

Low-motion talking-head sequences and high-motion action sequences represent the two ends of the interframe redundancy spectrum. In other words, low-motion videos such as talking-head shots have a lot of interframe redundancy, while high-motion video has very little. In low-motion videos, individual frame quality is critical, because the eye can focus and judge quality. In high-motion sequences, display rate is critical because the choppiness caused by slower frame rates is more obvious than a lack of individual frame quality. For this reason, it's important to review both types of sequences when selecting a codec.

We captured all the video sequences in technology-neutral RGB format. All are compressed to under 150 kB/second, representing the bandwidth of a single-speed CD-ROM drive.

After reviewing the results from the three real-world videos, we'll examine a video converted from an Autodesk animated FLI file. Many publishers find this conversion helpful because animated files don't support audio/video interleaving, making sound synchronization from a CD-ROM virtually impossible. As we've seen, the "born-to-interleave" AVI file format make sound synchronization a snap.

Other Considerations

Palette Management—When your target platform sports 8-bit video, you have to take special care to avoid "flashing," which occurs when a video's palette is different than the current display palette. Codecs handle palettes differently, and some work better with 8-bit displays than others.

Smaller Video Sizes—Often you'll want to use smaller video resolutions such as 240x180 for certain presentations. However, some codecs are optimized for 320x240 videos and don't perform well at smaller sizes. This causes a surprising disparity in video quality at the smaller resolutions.

The Contenders

 In Figures 6.1–6.12 we tested Cinepak version 1.5.0.198, Indeo 3.2 beta version 3.22.01.22, Video 1 version 1.0 and MPEG version 1.1. Microsoft's RLE had no version.

Low-Motion Sequences

Figure 6.1 Original

Figure 6.2 Cinepak—solid appearance, not as sharp as Indeo

Figure 6.3 Indeo—low-motion king!

Figure 6.4 MPEG—8-bit MPEG causes the obvious banding and QSIF implementation causes the fuzziness

Figure 6.5 RLE—while technically the only lossless codec, RLE shows severe loss even in low-motion shots

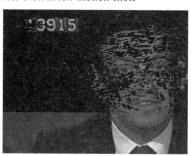

Figure 6.6 Video 1—good quality in low-motion sequences

High-Motion Sequences

Figure 6.7 Original—shot of a roller coaster sequence . . .

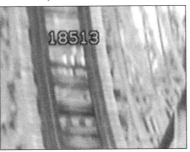

Figure 6.8 Cinepak—still the high-motion champ, beating Indeo 3.2

Figure 6.9 Indeo—version 3.2 made great strides, but fuzziness and artifacts still very apparent

Figure 6.10 MPEG—motion doesn't really bother MPEG, since all frames are key frames. Still fuzzy, though

Figure 6.11 RLE—lossless?

Figure 6.12 Video 1—MotiVE technology breaks up in high motion

Display Rate

We measured display rates with VCS Play. Table 6.1 contains our results. Because MPEG is not a Video for Windows codec, we couldn't test frame rate on VCS, and Xing doesn't offer a utility to capture frame rate. On a subjective basis however, Xing seemed the fastest codec by far, displaying all frames with apparent ease.

We included results of Indeo 3.1 to show how significant the new beta release of 3.2 appears to be. It's the first time Indeo matched Cinepak in display rate, which is critical for performance on high-motion files.

Table 6.2 shows that Cinepak retains a decompression speed advantage for 30 fps files. This obviously will translate to faster performance on 15 fps files on machines slower than our test bed.

Other Considerations

Palette Management—Most Windows-based video graphics cards display in 8-bit, 16-bit or 24-bit color depth, with 8-bit systems predominating. As we've seen, most codecs work

Table 6.1 Display rate by code and sequence

Codec Clip	Cinepak	Indeo 3.2	Indeo 3.1	RLE	MPEG	Video1 (8-bit)
8-bit						
Action	15	15	12	15	NA	15
Talking Head	15	15	14	15	NA	15
16-bit						
Action	15	15	9	15	NA	15
Talking Head	15	15	10	15	NA	15
24-bit						
Action	15	15	10	15	NA	15
Talking Head	15	15	11	15	NA	15

Table 6.2 High-end testing for Cinepak and Indeo 3.2

30 fps high motion file	Cinepak	Indeo 3.2	Indeo 3.1
8-bit display	30 fps	30 fps	30 fps
16-bit display	30 fps	16 fps	10 fps
24-bit display	30 fps	20 fps	23 fps

faster in 8-bit mode—especially native 24-bit codecs like Indeo and Cinepak. This is because there's less data to display to the Windows buffer and less to pump from main memory to the video card. This makes 8-bit mode very valuable when working with lower-end computers such as ISA bus machines.

When a computer is in 8-bit graphics mode, display is limited to 256 colors. This collection of 256 colors is called the palette, since all screen elements must be painted with colors contained in the palette. The 256-color combination is not fixed—palettes can and do frequently change. But at any one point, only 256 colors can be used to describe all the objects on the screen.

When displaying in 8-bit mode, all codecs are limited to 256 colors. For 8-bit codecs (Video 1, RLE, MPEG) this isn't a problem—their video is already described in 256 colors or less. This means that 8-bit codecs look as good in 8-bit mode as they do in the higher color depths. Conversely, this means that 8-bit codecs don't improve in higher color depths.

Cinepak and Indeo are native 24-bit codecs that use over 16 million colors to describe their compressed video. When displaying in 8-bit mode, Cinepak and Indeo have to drop from 16 million to 256 colors. To conserve file size and preserve display rate, neither codec stores palette information in their file—they simply decompress to the same fixed palette for all videos.

To minimize the potential for distortion, both codecs "dither" or draw minute geometric pixel patterns of various sizes to simulate colors not contained in the palette. Dithering is shown in Figure 6.13.

Figure 6.13 A dithered Indeo file

Palette Flashing

As noted earlier, all 8-bit displays use a set palette—as soon as you boot Windows in 8-bit mode, the palette is set. All graphic objects—animation, video, bit-mapped and/or vector image—have their own palette. When a graphic object displays, Windows makes certain that its palette is installed. If it isn't, Windows automatically changes the palette to that of the new object by momentarily blanking out and "realizing" the colors of the new palette. Often called palette flashing, this effect can be quite distracting.

8-bit codecs can "hold a palette," which allows developers to select one palette for a screen or presentation, and compress all videos and other graphics to that palette. Cinepak and Indeo can't hold a palette. We'll discuss palette work-arounds that avoid flashing in Chapter 13. However, even if you don't flash, Indeo and Cinepak videos will dither, which is less attractive than nondithered video.

For this reason, many developers consider Video 1 for mass-market multimedia products that will probably play on lower-end machines with 8-bit displays. As we've seen, Video 1 simply doesn't have the horsepower for action footage. On the other hand, for low-motion sequences and, as we'll see in a moment, animated footage, Video 1 does a great job.

Animated Footage

As we discussed, many developers convert animated files to AVI format to enable sound synchronization. Most animated files fall in the extremely "low-motion" range, since usually the background is fixed and only certain objects in the foreground move.

Note, however that animations use fairly compact vector formats, while AVI is a bit-mapped format. Often when you convert an animation to AVI format you bump the data rate over CD-ROM rates. When this occurs, you typically have three choices: you can slow the animation, make it smaller or forget the AVI conversion.

We'll detail how to convert animations to AVI format in Chapter 12. Here we'll analyze which codec to use when converting (Figs. 6.14–6.19). MPEG isn't shown because we weren't able to successfully convert any animated sequences to MPEG format.

Smaller Video Resolutions

Smaller screen resolutions such as 240x180 are often preferable to 320x240 resolutions. In training videos, for example, smaller displays offer faster playback and additional space for other screen elements such as graphics and text.

Certain codecs—most notably Cinepak—appear tuned specifically for 320x240 displays, and quality degrades at the smaller resolutions. In contrast, Indeo performs well on low-motion videos at virtually all resolutions (Figs. 6.20–6.24). Once again, MPEG isn't shown because it doesn't compress at resolutions other than 320x240.

Animated Sequences

Figure 6.14 Original animation

Figure 6.15 Cinepak—fuzzy, smeared, and clearly not the best option

Figure 6.16 Indeo—shows its lossy roots with a fuzzy image

Figure 6.17 RLE—lossless?

Figure 6.18 Video 1—a solid choice, and holds a palette, too! This is a delta frame

Figure 6.19 Video 1—on a key frame. When working with animations, use as few key frames as possible—like one key frame for the entire video.

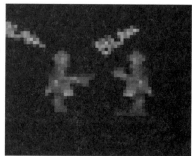

240×180 Videos

Figure 6.20 Original

Figure 6.21 Cinepak—note choppy mottled appearance

Figure 6.22 Indeo—offers highest-quality low-motion sequences over a range of resolutions

Figure 6.23 RLE—mottled and a bit fuzzy

Figure 6.24 Video 1—not quite as clear as Indeo, but not a bad choice

Performance Summary

High-Motion Footage—Overall, from a pure performance standpoint, Cinepak retains a slim lead over Indeo 3.2. At similar bandwidths, Cinepak provides the best overall quality and fastest decompression of the group. However, in our tests, the beta release of Indeo 3.2 indicates that Intel gained a lot of ground on Cinepak and now can seriously be considered for action footage.

Low-Motion Footage—Indeo shone in low-motion sequences, where its video quality was clearly preferable to that of Cinepak. With the speed increase delivered by version 3.2, using Indeo is now relatively pain-free from a display-rate perspective. We also preferred Indeo's palette handling in 8-bit mode over Cinepak's.

However, always consider Video 1 when working with low-motion videos targeted toward 8-bit display platforms. In the 8-bit environment, Video 1 delivers a non-dithered image, which is always preferable to dithered. Because you can compress to a fixed palette with Video 1, the 8-bit codec is also much easier to work with than either Cinepak and Indeo from a programming perspective. Later we'll look at a mode of Indeo 3.1 called Quick Compressor. While it only works in limited circumstances, one of them is low motion videos.

Smaller Video Resolutions—Indeo is best choice when working with resolutions smaller than 320x240. Time after time when we tried to use either Cinepak or Video 1 for smaller video resolutions, the results were mottled and distorted. Other developers have reported the same results.

Animated Sequences—When converting animated sequences to AVI files, try Video 1 first. Remember to use only one key frame for the entire sequence if you can. If Video 1 doesn't provide acceptable results, try RLE and Indeo 3.2 in that order.

Figure 6.25 A video captured with ScrnCap, the Video for Windows 1.1 screen capture application

```
┌─────────────────────────────────────────────────────┐
│ ─   VidEdit - MYFILE.AVI [Frame 599]        ▼  ▲     │
│   File   Edit   View   Video   Help                   │
│  ┌──────────────────────────────────────────────┐    │
│  │ ─          Compression Options                │    │
│  │  Target:                         ┌──────────┐ │    │
│  │  ┌────────────────────────────┐  │    OK    │ │    │
│  │  │ Custom                   ±│  └──────────┘ │    │
│  │  └────────────────────────────┘  ┌──────────┐ │    │
│  │  Video Compression Method:       │  Cancel  │ │    │
│  │  ┌────────────────────────────┐  └──────────┘ │    │
│  │  │ Intel Indeo(TM) Video R3.1 ±│              │    │
│  │  └────────────────────────────┘               │    │
│  │  ┌──────────────┐ ┌─────────────┐ ┌─────────┐ │    │
│  │  │ Save as Default│ │ Use Default │ │Preview>>│ │  │
│  │  └──────────────┘ └─────────────┘ └─────────┘ │    │
│  │  ──────────────────────────────────────────── │    │
│  │  ☒ Data rate              261  ▲ KB/sec        │    │
│  │                                ▼               │    │
│  │  ☒ Interleave audio every  1   ▲ Frames        │    │
│  │                                ▼               │    │
│  │  ☒ Key frame every        15   ▲ Frames        │    │
│  │                                ▼               │    │
│  │  ☐ Pad frames for CD-ROM playback              │    │
│  └──────────────────────────────────────────────┘    │
└─────────────────────────────────────────────────────┘
```

Screen Captures—One video category that we didn't review was screen captures. In Chapter 3 we discussed ScrnCap, a new screen capture utility introduced in Video for Windows 1.1. ScrnCap lets you capture sequences of screen commands such as dragging a mouse to open a dialog box, or highlighting a key control. You can add synchronized audio to prepare a multimedia training video like that shown in Figure 6.25. All ScrnCap videos are stored in RLE format, and in this one application, RLE truly lives up to its lossless heritage.

What about Xing?—Notwithstanding Xing's impressive decompression speeds, the fuzziness caused by its QSIF resolution scheme detracts from image quality. In addition, Xing doesn't interleave video and audio, which limits performance from fixed-bandwidth playback devices such as CD-ROMs. If you've played back any MPEG files, you've probably noticed that the video starts and stops and that the audio breaks up. However, Xing is the only codec offering real-time capture to very low bandwidths. This makes it an attractive alternative in certain applications, like security, that requires this feature.

MORE THAN YOU EVER WANTED TO KNOW

Now that we've seen how the codecs perform, we'll take a look "under the hood" using VCS Play's frame profile feature. As we've discussed, the frame profile separates each frame into its audio and video components, calculates individual frame size and creates a histogram of the results. We'll use the frame profile with some specially created video files to examine how the codecs control bandwidth, measure their interframe efficiency, determine how they manage major transitions and examine the trade-offs made to achieve their performance goals.

This helps us understand why the codecs perform as they do, and avenues for potential improvement. Since Video 1 and RLE have well-defined, niche uses, we'll limit our examination to Cinepak and Indeo 3.2.

Compression Management

Let's start by looking at how the individual codecs manage the compression process. Figure 6.26 contains frame profiles for the high-motion files shown previously (Figs. 6.8 and 6.9), with Indeo on the left and Cinepak on the right. The black

Figure 6.26 Frame profile for Indeo 3.2 (left) and Cinepak (right). Solid wavy line is per second data rate compared to the 150 kB/s and 300 kB/s limits. Protruding frames on right are key frames. Indeo's key frames are not distinct

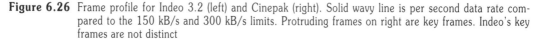
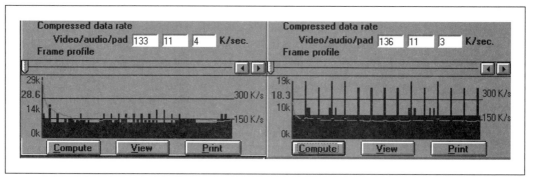

areas with small spikes are the individual frames. As we've seen, the wavy line is the per second data rate which shows whether a video file will play from a CD-ROM. The compressed data rates, separated into video, audio and CD-ROM padding, are shown above the graph.

Note that the two vertical axes are using different scales. The Indeo file on the left has an upper limit of 29 kB, where Cinepak's scale on the right peaks at 19 kB. This reflects our decision to optimize frame profile display for the individual file rather than for comparison between files. This provides the optimum view of each particular file, with relevant control information provided through the compressed data rate calculations.

Some review will assist in your understanding of the frame profiles. Cinepak and Indeo both use a system of interframe and intraframe compression that divides the video stream into two types of frames, key and delta frames. Key frames are frames upon which no interframe compression is performed. These provide a reference for subsequent delta frames and a point of access to the video. Typically larger than delta frames, key frames are easily identified as the tall frames in the frame profile on the right.

Delta frames are frames that exclude redundant information from previous frames, creating the interframe compression. Typically smaller than key frames, delta frames are the smaller frames in the previously identified frame profile.

We'll cover CD-ROM padding in detail in Chapter 12. For now, CD-ROM padding is filler bits placed in individual video frames to boost frame size to a multiple of 2 kB, the size of storage sectors on a CD-ROM. This forces all video frames to start and stop on a sector boundary, allowing the CD-ROM to stream data from the disk during retrieval rather than performing multiple seeks. For example If the video and audio components of a frame equaled 7 kB, you would add 1 kB of padding to round the frame to 8 kB. Obviously, these garbage bits add nothing to video quality, so a codec's ability to minimize CD-ROM padding is key to getting the most out of your fixed video bandwidth.

Figure 6.26 shows that at 15 frames per second, Cinepak and Indeo are fairly similar. Cinepak obviously retains a clearer distinction between key and delta frames and greater individual control over delta frame sizes. However, both codecs manage CD-ROM padding equally well, with Indeo producing 4 kB/second of this filler data, and Cinepak 3 kB/second.

Figure 6.27, illustrating the same high-motion video compressed with Indeo 3.1, shows just how far Indeo 3.2 has come. Version 3.1 did not optimize CD-ROM padding. The optimization in Indeo 3.2 allows Intel to deliver about 12 kB/second more video (133 kB/s as compared to 116 kB/s) in this 150 kB file, which translates to a quality improvement of about 10%.

Indeo 3.1 also used a rather crude compression management option when up against the 150 kB/second bandwidth limit—they simply dropped a frame during compression and duplicated the same frame twice. These drop frames are the valleys in frames 13 and 28. If you load Fig6_27.avi, and step to these frames, you'll see that they're identical to the frames immediately before them. Improved compression management in Indeo 3.2 enables Intel to avoid dropping frames at 15 fps. As we'll see in a moment, however, Indeo 3.2 still resorts to this approach when really pushed against a bandwidth limit.

Figure 6.27 Indeo 3.1 frame profile of same high-motion video file. Valleys are frames dropped during compression

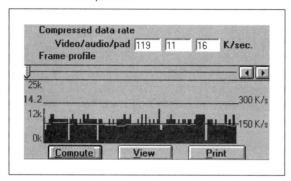

Half as Much Is Twice as Good

Figure 6.28 Original frame

Figure 6.29 Four frames later in sequence

Figure 6.30 Cinepak—note smooth arms even though image is zoomed 2x

Figure 6.31 Cinepak stressed, and clearly dealing with pixel blocks, not single pixels

Figure 6.32 Indeo 3.2—note smooth arms

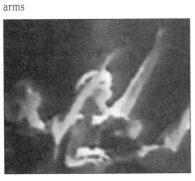

Figure 6.33 Indeo 3.2 stressed. Note blocky effect similar to Cinepak

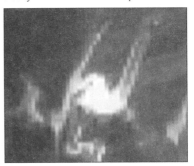

Figure 6.34 Indeo 3.1—less crisp than V 3.2

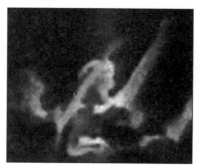

Figure 6.35 Indeo 3.1 stressed. Compare middle arm with Figure 6.33

How did Indeo produce the performance boost in version 3.2? Well, at least in part, it looks like they did it the old-fashioned way—by imitating their competitor. One technique used extensively by Cinepak is to work with bigger blocks of pixels during periods of extremely high interframe motion. This is often referred to as subsampling.

Essentially, in high-motion periods, Cinepak subsamples the 320x240 frame down to 160x120 and compresses the frame at that resolution. Since it contains four times less data, it's four times easier to compress. After compression, Cinepak pixel-replicates the video frame back out to 320x240, causing the blockiness shown in Figure 6.31. This technique works well because in high-motion sequences it's difficult to notice occasional 2:1 pixel-replication artifacts.

Compare Indeo 3.2 (Figure 6.33) to Cinepak (Figure 6.31) and you'll notice that Indeo 3.2 uses the same technique. In fact, the middle arm looks virtually identical in the two pictures. However, Indeo appears to subsample less, creating a clearer frame. If you check back to Indeo 3.1 (Figure 6.35), you'll see that the older version either didn't use this technique at all, or was much more selective. Obviously, from Intel's perspective, a little pixel replication is preferable to wholesale frame dropping during compression. But, as we'll see in a moment, old habits die hard.

Old Habits Are Hard to Break

What happens when Indeo gets really stressed? We compressed a 30 fps action sequence with Indeo 3.2 and Cinepak to find out. The results are shown in Figure 6.36.

As you can see, even Indeo 3.2 starts to drop frames during compression during high-motion sequences. However, since for the most part we recommend that you work with 15 fps files, this shouldn't cause a problem. As we've said, Indeo 3.2 is substantially more powerful than Indeo 3.1—this just shows that Indeo has further room for improvement.

Interframe Efficiency

Interframe efficiency refers to two performance characteristics, how quickly the codec recognizes interframe redundancies and how compactly it updates subsequent frames for this redundant information. To measure this, we used an artificial video file created by duplicating a simple bitmap 30 times. (See Figure 6.37.)

Cinepak, on the right in Figure 6.38, immediately recognizes that the frames are identical and reduces delta frames to about

Figure 6.36 Frame profile for 30 fps video compressed with Indeo (left) and Cinepak (right). In this high-motion sequence, even the new version of Indeo drops frames

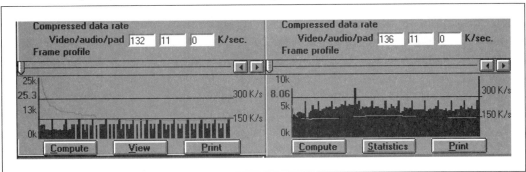

Figure 6.37 Artificial video consisted of 30 identical frames like this one

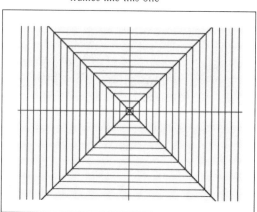

800 bytes after each key frame. Indeo takes two frames after each key frame to recognize that the frames are identical and then reduces delta frames to about 0.76 kB, just under Cinepak. Overall, in this test Cinepak outperforms Indeo by about 6 kB/second, which is certainly significant, but not substantial. Although a separate frame profile is not shown here, Cinepak outperformed Indeo 3.1 in this test by about 10 kB/second.

Figure 6.38 Frame profile for Indeo (left) and Cinepak for video comprised of consecutive identical frames. Key frame setting for both codecs is 15, which accounts for the "spike" halfway through the video

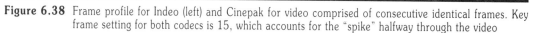

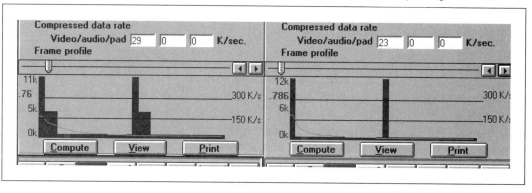

Transition Management

Transitions are abrupt changes in video content, such as scene changes. Transition frames are difficult for codecs to manage because little or no interframe compression is available. To test how the codecs managed transition frames, we created another 30-frame sequence, comprised of 15 frames from the previous line drawing video, and 15 identical frames from a talking-head video (see Figure 6.41).

We compressed twice. The first test, shown in Figure 6.39, illustrates how the codecs manage transitions occurring on a key frame. Key frame setting for both codecs in this test was 15, which corresponded to the transition. The second test, shown in Figure 6.40, illustrates how the codecs react to transitions occurring on delta frames. The key frame settings for this test was 10.

As shown in Fig. 6.39, when the transition occurs on the key frame, both codecs managed the transition very effectively. While Indeo took three frames to learn that the frames were identical, both in the beginning and at the transition, both its key and delta frames were smaller than Cinepak's. For this reason, Indeo 3.2 matched Cinepak's data rate. In contrast, Indeo 3.1 produced a data rate of 50 kB/second, which is about 42% higher than Cinepak.

When the transition occurred on a delta frame, the two codecs reacted differently. Cinepak *created* a key frame, which allowed it to manage the transition efficiently. The large frame in the middle of the right-hand table is the key frame. After creating the key frame it resumed the requested key frame interval of 10.

Indeo created two delta frames the approximate size of a key frame, which appear as the two large frames in the middle of the right-hand table. It then placed a key frame four frames later, ignoring the benefit provided by the two large delta frames. However, because Indeo's key frames were smaller than Cinepak's, overall its data rate exceeded Cinepak's by only 5 kB/second, or 10%. This is insignificant considering that Indeo produced a higher-quality video than Cinepak (see Figures 6.41 and 6.42). For the record, Indeo 3.1 produced a

Figure 6.39 Frame profile for Indeo (left) and Cinepak for artificial video with transition on a key frame

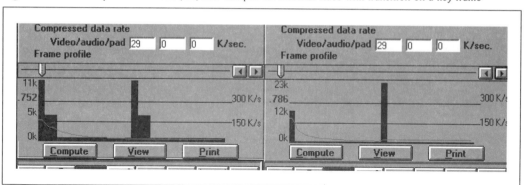

Figure 6.40 Frame profile for Indeo (left) and Cinepak for artificial video with transition on a delta frame

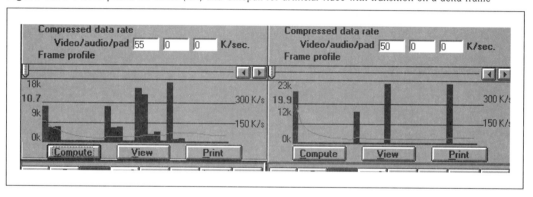

Figure 6.41 Indeo 3.2 producing crisp, clear video

Figure 6.42 Cinepak producing mottled, fuzzy video

data rate of 78 kB/second, which was close to 60% less efficient than Cinepak.

Summary

Overall, Cinepak is both more efficient and more innovative than Indeo, which is probably to be expected, given its roots in the software-only playback arena. From the start, Cinepak was designed for efficient playback without hardware assistance, and it performs accordingly.

Indeo is a codec in transition. Its roots are in DVI, which is capable of 30 frames per second of broadcast-quality video with hardware decompression. However, to convert DVI to software-only playback required significant retrenchments in compression features. In essence, Intel is approaching the market from the opposite side of Cinepak and must adjust accordingly.

However, somewhere in the bowels of Intel, someone knows how to create broadcast-quality video. Indeo Version 3.2 clearly evidences that. With advances in processor and bus technologies, standards such as DCI and new technologies such as video co-processors, the power behind the video surpasses that of the old DVI hardware. Indeo is slowly returning to its hardware roots, and its hardware performance won't be far behind.

Intel is much larger than SuperMac and has more at stake with Indeo than SuperMac does with Cinepak. In addition to serving as the core technology for Intel's Smart Video Recorder, Indeo is also featured in Intel's new Pro Share video-conferencing system, which has a market potential that far exceeds the Smart Video Recorder. Pro Share also fulfills Intel CEO Andy Grove's promise of video on the desktop.

SuperMac, recently purchased by Radius, appears to be in retrenching mode. They consolidated operations and recently lost most of Cinepak's early developers, who refused to leave Oregon for Southern California.

Overall, Indeo 3.2 is a very significant release for Intel, which erases much of Cinepak's quality and speed advantages.

It's a very clear sign of Intel's commitment to Indeo. Version 4.0, due out in 1995, could be the first to leapfrog Cinepak in performance.

SuperMac won't be sitting still, but Indeo has more potential areas for improvement and more incentive. Stay tuned.

LICENSING MECHANICS

Sold on Video for Windows? Here's how you can distribute the videos.

The following information is excerpted from Video for Windows 1.1 Release Notes.

> Your license agreement for Microsoft Video for Windows allows you to create AVI sequences and distribute them along with the Media Player so that others can view those sequences.
>
> Microsoft grants to you the right to reproduce and distribute the runtime modules of the SOFTWARE provided that you:
>
> (a) distribute the runtime modules only in conjunction with and as a part of an application program or data file, and not as part of any operating system or utility program, or on a standalone basis;
>
> (b) do not use Microsoft's name, logo, or trademarks in any marketing or advertising;
>
> (c) include a valid copyright notice on your product; and
>
> (d) agree to indemnify, hold harmless, and defend Microsoft and its suppliers from and against any claims or lawsuits, including attorneys' fees, that arise or result from the use or distribution of your product which incorporates the runtime modules.
>
> The license in this section to distribute the runtime modules is royalty-free provided that your application program or data file incorporating the runtime modules is created for operation on the Microsoft(R) Windows(TM),

Windows NT (TM), or Modular Windows (TM) operating systems.

All of the Video for Windows codecs reviewed in this chapter are included in the runtime, so within these guidelines you're free to distribute the codecs—both compression and decompression capabilities—as your heart desires. Note that the runtime does not include utility programs such as VidEdit and VidCap, so you can't distribute these programs without making special arrangements with Microsoft. Xing Technologies also has a royalty-free runtime policy.

You should probably assume that all other codecs will not follow this policy and will charge for runtimes.

Runtime Mechanics

You have to install the runtime to play AVI files, so you should plan to ship the runtime with your video files. Luckily, it all fits on one 1.44 MB floppy.

Typically you'll acquire Video for Windows with a capture board or authoring system. Many of these products collect dust for months before finding their happy home in your computer. A buddy recently (June, 1994) picked up an Intel Smart Video recorder which came with Video for Windows version 1.0—only about five months out of date. Intel gladly sent him version 1.1.

Microsoft, Intel and SuperMac all freely distribute their latest runtimes on Compuserve. You'll find them in the Computers and Technology Software Forums. You should make a habit of checking Compuserve before shipping to learn about any new versions, bugs or enhancements. Nothing's worse than shipping known bugs or old versions.

I know. Trust me.

The Video for Windows runtime includes Media Player, to provide basic video playback capabilities, and a setup program that you're also free to distribute. However, the setup program is pretty basic and does not check for newer versions of files before installing. For example, if your new customer just downloaded the latest Cinepak release from Compuserve, and then installs your product, Microsoft's installation program will write right over it, which isn't the best way to start a relationship with a customer. You might consider writing your own installation program with the information supplied in the Video for Windows release notes.

TOUR SUMMARY

1. Selecting a codec isn't like marriage—you don't pick one for life. Instead, get familiar with their respective strengths and weaknesses, evaluate the video and primary target platform market and pick the best codec for the job. Most multimedia publishers use two or three different codecs within their titles.

2. Here's our guide.

   (a) *Low-motion videos*—Indeo, unless 8-bit target market, then consider Video 1.

   (b) *High-motion videos*—Indeo 3.2 is quickly catching up, but for now Cinepak is the best choice unless . . .

   (c) *Working with resolutions of less than 320x240*—you have to consider Indeo 3.2.

   (d) *Animation*—Video 1, RLE, Indeo, in that order.

3. To paraphrase an old cigarette advertisement, "Indeo 3.2 has come a long way, baby." If your last experience with Indeo was with version 3.1 or earlier versions, 3.2 deserves another look.

4. All Video for Windows codecs and the runtime itself are royalty-free for most uses. Microsoft even supplies the installation program.

ANALOG
OVERVIEW

ANALOG OVERVIEW

IN THIS CHAPTER

All digital video starts out as an analog signal on film, and a working knowledge of analog standards and formats is essential to successfully navigating the analog-to-digital conversion. This section starts with an overview of analog storage formats such as SVHS, BetaSP and Hi-8, and then moves on to review broadcast, signal and other analog standards.

We'll finish with the Great Analog Source test, where we examine the effects of these analog characteristics on ultimate compressed video quality. The answers will affect both how you film and how you capture video.

ANALOG STANDARDS

Let's jump right in and attack the maze of analog standards using Fig. 7.1 as a guide.

CAPTURE/STORAGE FORMATS

Whenever you film or edit, you store the video in a particular analog video format, represented in the upper left-hand corner of Fig. 7.1. At a minimum, you're probably familiar with VHS, SVHS and Hi-8. As we'll see in a few moments, differences

Figure 7.1 Analog tape, broadcast and signal standards

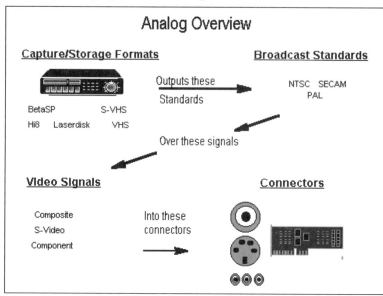

between the formats relate to issues such as how color and black-and-white information is stored, and signal bandwidth.

The highest-quality format shown in Fig. 7.1 is BetaSP. While there are higher-quality formats, such as one-inch, D-1 and D-2, costs for components handling these formats are sky-high. In contrast, while BetaSP cameras and decks are out of the purchase price range for many developers (Doceo included), these components are widely available for rental at reasonable rates. For this reason, we included them in our survey and testing.

BROADCAST STANDARDS

Broadcast standards, in the upper right-hand corner of Fig. 7.1, define how the signal is represented when transmitted to a receiving device such as a television. NTSC, which stands for National Television Standards Committee, is the reigning standard in North and Central America and Japan. We'll look at the background and color composition of NTSC in a moment. For now, remember that NTSC signals define the screen with 525 lines per frame, and update the screen at just under 30 frames per second.

PAL, for Phase Alternation Line, dominates in Europe, and is also found in the Middle East, Africa and South America. SECAM, for Systemem Electronic Pour Couleur Avec Memoire, is a PAL variant used in France, Russia and other pockets in Africa. Both PAL and SECAM use 625 lines of resolution and 25 frames per second.

For the most part, VCRs, decks and other analog players handle only one format. Fortunately, if you're working in the United States, you'll rarely run into PAL or SECAM materials, and I assume the inverse is true in other geographical areas. If you get handed a PAL or SECAM tape, check your yellow pages under Video Production Services and start making phone calls. I'm sure someone local can either convert the tape to NTSC or recommend a regional resource that can do the work.

VIDEO SIGNAL STANDARDS

Video Signal Standards, in the lower left-hand corner, relate to how color and other video information is stored and transmitted. The transmission method also impacts the type of connector used on the video capture card, shown in the bottom right-hand side of Fig. 7.1. A composite signal, which carries all information in one channel, uses a one-hole jack called the RCA Phono connector. The S-Video signal, composed of two channels, uses a four-pin connector called the mini-DIN connector. A true component signal, made up of three separate video signals, hooks into three separate connectors.

The key concept to remember is that all U.S. video decks record and play, and all U.S. capture boards receive, an NTSC signal. The signal may be stored on a laserdisc or BetaSP deck, and may be transmitted as composite, S-Video or Component video. However, all these formats are under the umbrella of NTSC video.

The Evolution of Analog Signals

It all started with television—but you probably already guessed that. Back in 1953, the NTSC committee had a problem. Black-and-white televisions were pervasive, but color was coming. They had to design a standard that would work with both.

Black-and-white televisions receive one signal, called luminance, which is often referred to as the "Y" signal. Each screen pixel is defined as some range of intensity between white (total intensity) and black (no intensity). To maintain compatibility with older black-and-white sets, the NSTC had to set a color standard that kept the luminance signal separate and also provided the color information required for newer color television sets.

In the digital world, we're comfortable with colors described with red, green and blue, or RGB, values. Your first color monitor probably said RGB right on the case. The analog world has also embraced the RGB standard, at least on the acquisition side, where most cameras also break the analog signal into RGB components.

However, the NTSC couldn't use RGB as the color television standard because the old black-and-white television sets couldn't decode an RGB signal. They had to send a luminance signal for the black and white sets, and fill in the color information with other signals, called hue and saturation, or U and V. For this reason, where the digital world works around RGB, the analog world, especially around television broadcasting, works in YUV.

Component, S-Video and Composite

Figure 7.2 traces the evolution of the analog signal from RGB to composite. On the extreme left-hand side of the chart is RGB capture. Here, storage channels are maintained for each of the primary colors.

However, RGB is an inefficient analog video storage format for two reasons. First, to use RGB, all three color signals must have equal bandwidth in the system, which is often inefficient

Figure 7.2 The evolution of an analog signal from RGB to NTSC

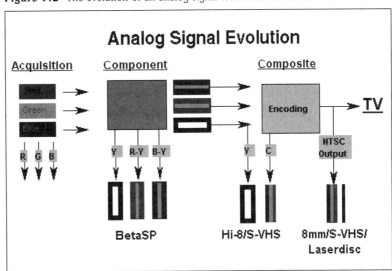

from a system design perspective. Second, since each pixel is the sum of red, green and blue values, modifying the pixel forces you to adjust all three values. In contrast, when images are stored as luminance and color formats, as in the YUV format, you can modify a pixel by changing only one of the values. For this reason, the Y, R-Y and B-Y shown in the middle of the chart was adopted by CCIR601 as the international component video standard.

Component video means that separate channels are maintained for each color value, both in the recording device and on the storage medium. This minimizes noise that occurs when two signals are combined in one channel.

After NTSC encoding, the hue and saturation channels (U&V) are combined into one chrominance channel, also called the C channel. A video signal called S-Video carries separate channels for the luminance (Y) and chrominance (C) signals. This is also called Y/C video.

To play on our old black-and-white televisions, we know that all color and other information must be ultimately combined into one YUV signal, called a composite signal. This is represented on the extreme right-hand side of the screen.

Technically, a composite signal is any signal that contains all the information necessary to play the video. In contrast, any one individual channel of component or Y/C video would not be sufficient to play the video.

Capture and Storage Formats— The Contenders

As you can see from Fig. 7.2, the first major difference between the capture formats is how the color information is stored. BetaSP stores the color information as component video with three separate channels. Hi-8 and SVHS uses the two-channel Y/C video, while lowly VHS, 8mm and laserdiscs store all color information in one channel.

Which signal produces the best video quality? The composite signal combines all signals into one channel, which inevitably introduces noise into the signal. Noise is manifested, in the terms of one extremely technical manual on the subject, as "any random fleck that shows up in the display." I couldn't have said it better myself. So obviously the composite signal is the noisiest.

The S-Video signal (also Y/C or separate video) produces less noise, since the two signals are isolated in separate channels, not merged together. This minimizes flicker and color blur. Finally, the component signals provides the highest-quality signal, since all components are maintained in separate channels.

To preserve this color fidelity, you should transmit the video to the capture card in as separate a video signal as possible. This relates back to the video signals and connectors shown in Fig. 7.1. For example, virtually all BetaSP decks have composite out signals, which combine the three component signals into one channel. This obviously negates some of the benefits of using BetaSP, and component or S-Video signals are preferred.

Unfortunately, the inverse is also true. For example, SVHS decks usually play VHS tapes as well. Broadcasting a VHS signal through the S-Video cable doesn't reverse noise that may

Table 7.1 Analog capture and storage standards

	BetaSP	SVHS/Hi-8	VHS/8mm	Laserdisc
Color signal	Component	Y/C	Composite	Composite
Lines of resolution	360	400	240	400
Signal to noise (s/n) ratio	50 dB	47 dB	47 dB	47 dB

be present in the original signal. You may get a slight improvement in quality if the deck performs a cleaner color separation than the capture board, but you won't get SVHS quality.

One final point. Obviously a capture board is only as good as the signal that it can accept. Surprisingly, some higher-end capture boards only accept composite input signals. While this doesn't render the board totally useless, it does indicate that the board left some video quality "on the table" and isn't performing as well as it could.

The tape formats also differ in other characteristics, as shown in Table 7.1.

LINES OF RESOLUTION

How does lines of resolution affect video quality? During broadcast, all decks must convert the stored lines of resolution into the 525 lines required for the NTSC signal. VHS and 8mm formats have to convert the 240 lines of resolution to the 525 lines. More significant is the medium's signal to noise (s/n) ratio, which describes the amount of data in the signal, or signal strength. As you'd expect, the higher the signal to noise ratio the better. For example, a drop of 3 decibals (dB) translates to a 50 percent increase in noise.

Hi-8, laserdiscs, and SVHS capture and store 400 lines of information. While some interpolation occurs when outputting to the 525-line NTSC signal, it's only about 1.25:1 as compared to over 2.5:1 for VHS. This translates to accuracy and sharpness. Hi-8, laserdiscs, and SVHS capture and store video information at a s/n ratio of over 45 dB, which further enhances video quality.

Finally, BetaSP, the most popular professional format, has 40 fewer lines of resolution than Hi-8/SVHS, and its s/n ratio is 6 dB higher. All other things being equal, you're probably guessing that BetaSP is the preferred format, and you would be right. But if your name isn't Rockefeller, and you've checked the prices of BetaSP equipment lately, you're probably asking yourself—just how much does it help?

The answer is—it helps a lot, in certain places under certain circumstances. To see exactly where, let's trace the flow of analog video from the original filming to digitization.

FROM ANALOG TO AVI

Filming

This is the part where you buy makeup, Klieg lights, invite all your friends and shout things like "Quiet on the set," "Cut," and my personal favorite, "Print it." The fun part, before all the serious work begins.

Anyway, when you film, you have a range of options, from your personal Hi-8 camera to the local BetaSP deck that rents without operator for about $500 per day. Does choice of format matter here? Yeah, it does. Here's how we tested.

The Lens Test

We shot the same video with three cameras: a Sony 3-CCD camera storing in BetaSP format that cost around $25,000; a Canon A1 that cost around $3,000; and a Sony TR101 Hi-8 camera that cost around $1,100. We filmed at the GTE Visnet studios in Atlanta, where we shot all the footage for the *Video Compression Guide and Toolkit.*

Through GTE, we hired a professional camera operator for the Sony 3-CCD camera. He was also familiar with the Canon A1 and configured the camera to match the Sony 3-CCD as closely as possible. The Sony TR101 was configured with care by an associate who is a practiced video enthusiast out to prove that his TR101 could match the Canon costing three times more. He was motivated.

Rather than shoot from exactly the same camera angle in sequence, which would yield similar lighting but cuts that would be difficult to compare, we positioned the two Hi-8 cameras as closely as possible to the Sony 3-CCD camera and shot once. I felt like Michael Jordan the day he quit basketball. While we got as close in angle, height, position and focus as we could, I'll be the first to admit that the videos created were slightly different in many respects. Overall, however, I think the results were instructive.

As was our practice, we edited the "keeper" takes from the original BetaSP film onto a one-inch master. One-inch tape is a component format with a higher signal bandwidth than BetaSP. This one-inch tape was therefore "second generation."

While preparing the tests presented in this chapter, we transferred the clip from the one-inch master back to BetaSP (third generation), which was ultimately transferred to Hi-8 tape for capture and comparison to the other two cameras (fourth generation).

To explain, we transferred to Hi-8 format to isolate the impact of lens quality on ultimate video quality. Had we captured from the original BetaSP film or one-inch master, we couldn't be certain how much of the quality differential, if any, related to media and how much related to the cameras.

We captured all three Hi-8 tapes using a Sony CVD-100 and ISURPro board. We captured at 15 frames per second, 320x240, and compressed the video to about 150 kB/s with Indeo 3.1. The results are presented in Figures 7.3, 7.4, 7.5.

Subjectively, the gap between the Sony 3-CCD camera and the two Hi-8 cameras is much more dramatic than the difference between the two lower-end cameras. When you consider that the Sony 3-CCD footage was ultimately captured from a fourth-generation source, you get the strong feeling that it pays to invest in quality equipment during the original analog capture.

The videos are contained in the Chapter 7 subdirectory on the CD-ROM under Fig7_3.avi, Fig7_4.avi and Fig7_5.avi, and I urge you to play them. Here are our observations.

Sony 3-CCD

The most striking image from a warmth and depth standpoint. The high-end camera handled the lighting most smoothly, without the white foreheads and nose present in both other videos—even taking into account that this camera was directly in front and the others were off to the side. The gray back wall was the smoothest and least mottled of the three cameras. However, we were surprised to find some patches of green in the black coat, indicating some color slippage.

You may notice that this camera was focused on the subject, which left the back wall softly out of focus. This seems to be a fairly useful filming technique, since flat surfaces are the bane of most codecs. By filming the back wall out of focus and somewhat fuzzy, you limit the flat, static detail that the codec can turn into a boiling mass of motion. This promotes the perceived quality of the back wall.

Canon

Showed better color fidelity than the Sony TR101, producing a smooth back wall and crisp black coat texture without swatches of green. As you play the video, you'll notice that the frames seem to jump around a bit. This tape jitter could have been caused by the Sony capture deck, however, as we noticed it on several other step captures.

The Canon video was the most impressive from a clear compression perspective. At 126 kB/second, the image was clear and virtually artifact-free. We attribute this to the fact that we captured first-generation footage from this camera. While editing your many takes into one reel of final footage saves a lot of time, you may get the best results capturing from the original footage. Notwithstanding the higher color quality and bandwidth of the BetaSP format, the first-generation capture from

Figure 7.3 Video filmed with a high-end Sony 3CCD camera and dubbed to a Hi-8 tape for capture. Colors are rich, deep and smooth, although the author would like to point out that the camera added at least 10 imaginary pounds.

On the negative side, whether the result of lens or simply the three generations that passed between filming and capture, the Sony 3CCD image showed some artifacting upon zooming that wasn't present in the Canon video at 20 kB/second less.

Figure 7.4 Video filmed with the Canon A1 camera. Color quality was less than that produced by the Sony 3CCD camera, and overall the image lacked the Sony 3CCD warmth

On the other hand, the Canon image compressed down to a virtually artifact free 126 kB/second, and the back wall is the smoothest of the three. Like vegetables from the garden, the best captures may come directly from the original film.

Figure 7.5 Video captures with Sony TR101 camera. The Sony produced a very credible video that trailed the Canon in pure quality and the Sony 3CCD in warmth. Note, however, that the angle of the shoot is at least partially responsible for some of these problems.

The camera showed some muddling in the back wall, and mottled green patches on the coat. However, the facial colors were much more accurate and pleasing.

the Canon produced the clearest, most artifact-free video of the bunch.

SONY

Did not like the gray back wall, which becomes more apparent at higher magnifications. Also produced the most green patches in the black coat. However, the camera performed exceptionally well for an $1,100 device, producing facial colors that were subjectively superior to those from the Canon.

SUMMARY

However, this experiment proves two points. First, when filming for digitization, it pays to use a high-quality camera. Camera rental costs for the *Video Compression Guide and Toolkit* totaled about $1,500, a fraction of the overall production budget. Given the importance of video to our overall product, it was money well spent.

Cameras like the Sony TR 101 and Canon are absolutely essential to the overall production process, especially for rehearsals and proofs of concept. However, you'll produce a higher-quality result for final production with a higher-end device.

The second point is that BetaSP's high s/n ratio allows the signal to hold up over multiple generations. If you plan exotic transitions or special effects that require multiple generations, BetaSP may be absolutely required to retain quality.

On the other hand, we also noticed some degradation in video signal when comparing the first-generation Canon Hi-8 video with third-generation footage from the higher end camera. While it may be a real pain in the rear, you may want to eschew an edited master and capture directly from the original tapes.

Finally, what this test didn't establish is whether BetaSP produces superior results when used as the analog format for capture. Here we learned that filming with a higher-end camera produces better compressed video, and that BetaSP holds up remarkably well through multiple generations. Still to come is whether sending a BetaSP stream to the capture board produces appreciably better video than Hi-8, SVHS or even VHS and laserdisc.

IN SEARCH OF THE PERFECT WAVE

OK, I admit it. There is a CPA in my background, those early years of the 1980s when I was fresh out of grad school and still running on that pre-programmed "you have to be a professional" course set in high school and college. I bailed out of public accounting one step before the cold auditor debited me out, leaving surprisingly little residue save a couple of old club ties and the nagging feeling that there is order in the world, if one could only quantify it.

Naturally, when I started compressing video, I felt there was a THE RIGHT WAY to capture video. I left Iterated Systems secure in my knowledge that step frame capture from a laserdisc was THE best way.

Then I learned that laserdiscs use a composite signal, the analog equivalent of being a smoker amidst nonsmokers. I polled my publishing brethren and saw platinum titles captured from VHS with $300 Video Spigots. I learned of mythical analog formats such as BetaSP and million dollar capture stations that churned out perfect files in near real time.

So, although somewhat less romantic than the *Endless Summer's* search for the perfect wave, but only somewhat, I began my quest for the perfect capture system. Or maybe Diogenes is a better analogy.

I took my lamp, my surfboard and my trusty Gateway 2000 computer figuratively around the world, testing analog decks from Japan, capture boards from Europe and the United States, and a good old, made in the US of A, eight-cylinder with overhead cam supercomputer that converts analog to digital without ever touching a capture card.

And while I never saw the sunset over the Australian Barrier Reefs, or smelled the salty morning tide off Honolulu, I have seen perfect raw footage that takes your breath away and learned the truth of how to video capture. These bring their own form of spiritual harmony and peace.

Hmmm. I have been working too hard on this book.

Anyway, I'll start with a story that's also an excuse. I began my high-tech career in fax boards. Seemed like a great place to be back in 1987.

In developing our little product, called the JT Fax, we tested with exactly one fax machine—our Sharp office facsimile. You see, Group III facsimile is standard, and if you can talk with one system, you can talk with them all. In truth, we tested **real hard,** sent multipage faxes at all hours of the day and night, testing during storm and sunspot, and released our product with full knowledge that it would be the most compatible fax board ever introduced.

I hear you chuckling. Yup, didn't work with fully half the fax machines out there, including those made by market leader Ricoh. Well, it took a while, but we got things squared away, as much, I later learned, as any fax device vendor gets things squared away. Sometimes I'm amazed fax machines work at all.

Segue to seven years later. I design what I believe is the perfect test for analog formats and capture boards, have my tapes dubbed by one of the best facilities in Atlanta and my laserdisc mastered at a world renowned facility. I play them all from professional-quality equipment, and guess what?

The analog world is no better than the digital one. All the clips look different. Not enough to disguise our ultimate conclusions. Just enough to be . . . distracting.

We captured with seven or eight of the best capture boards in the world, took great pains to make the input signals look at similar as possible, and guess what? They all look different. Well, in truth, given time we could tweak and adjust and get the final products looking a little bit more similar. But it's really a diminishing returns kind of thing, and my editor is telling me it's time to shoot the author and ship the book.

So. When you look at the clips, try to ignore the facts that (a) they're all of me, (b) sometimes I look like a Martian. Try to judge the formats not on absolute color, because that can be adjusted, but on color fidelity, as in why are there green spots on his black jacket? Also focus on resolution—which formats look as if they delivered the most detailed information to the capture card.

When you look at capture cards, compare quality and color fidelity. All were captured from the same BetaSP tape, so artifacts and muddled colors relate strictly to the board.

I'll end with a story as well. I recently taught a two-day seminar at a wonderful facility in Raleigh/Durham. The instructor's control center was like Houston Control, with access to five computers, four analog sources and three monitors and a 13-foot wall panel. Halfway through the seminar, after switching out two computers incompatible with my capture card and reinstalling Video for Windows at least seven times, I had to laugh. "Why would anyone want to demo with more than one computer? One machine can mess you up—four computers can kill you."

It's the same with your capture system. When you settle on one format, one deck and one capture card, you can learn their strengths, faults, limits and idiosyncrasies and consistently produce high-quality output. Like a marriage, if you stick with one setup, you can really make it work.

USABLE QUALITY— CAPTURE TECHNOLOGIES

By now it's relatively clear that BetaSP is the highest-quality format that's reasonably accessible from a budget standpoint. Does that automatically make it the best capture format? Again, sometimes yes, and sometimes no. It all depends on your ability to access the quality.

Real-Time vs. Step-Frame Capture

Figure 7.6 illustrates what happens during a typical 320x240 capture at 15 frames per second. Note that computer is a VESA local bus computer, but that the capture card sits in an ISA slot. While data transfers to the graphics card at a speedy 33 MHz at 32-bits bandwidth, transfer from the ISA-ridden capture card runs at 8 MHz and 16-bits.

Figure 7.6 Capture bottleneck caused by ISA bus during real-time capture

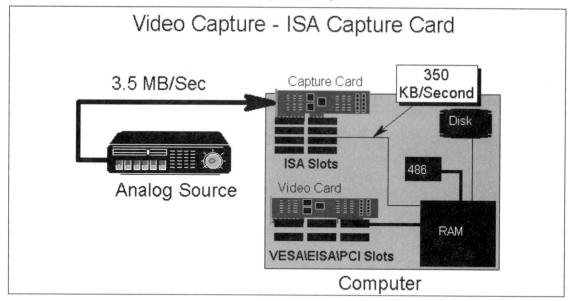

Data enters the capture card at about 3.25 MB/second, but the ISA bus extension will only support a sustained transfer of about 350 kB/second, about one-tenth the original stream. To capture in real time without dropping frames, the capture board must compress the video during capture, usually either with Motion JPEG or Indeo.

If your target data rate is under 200 kB/second, you'll probably have to recompress after capture to achieve your target. Since both compression steps are lossy, this means twice the loss—like photocopying a photocopy.

The alternative to real-time capture is called "step-frame" capture. In this mode, the capture software controls both the analog source and computer and feeds frames only when the capture board, bus and storage subsystem are ready for them, usually around a frame per second. Step frame enables compression-free capture without dropping frames.

To step-frame capture, the deck must have two characteristics. First, it must be frame accurate, which means that it can pause on a single whole frame and hold synchronization.

Second, it must be controllable through a communication standard supported by your capture software.

If you've ever "paused" your home VCR, you've probably seen that it isn't frame-accurate. Neither are any of the low-end BetaSP decks that I've experimented with. Which means that you can't practically step-frame in BetaSP format.

Until very recently, the only devices known to meet both criteria were laserdiscs, including the Pioneer CLD-V2600 we use at Doceo. However, in the last few months we learned of a Hi-8 deck, the Sony CVD-1000, that can also step-frame.

Looking back at Figure 7.2, we see that both Hi-8 and laserdisc formats are inferior to BetaSP. However, since both formats can step-frame, they can deliver the video without compression. BetaSP can't. So the real issue becomes—do step-framed Hi-8 and/or laserdisc formats deliver better quality than BetaSP compressed during capture?

Step Frame Vs. Real Time—Experiment

The test footage used here is also from the *Video Compression Guide and Toolkit*. Let's review the generations of the respective footage, which is all derived from the same video sequence filmed on BetaSP (Table 7.2).

As you can see, the laserdisc was one generation newer than the other two. However, since all previous generations were

Table 7.2 Generations of test formats

Tested Format	1rst Generation	2nd Generation	3rd Generation	4th Generation
BetaSP	Filmed on BetaSP	Dubbed to one inch	Dubbed to BetaSP	Dubbed to BetaSP
Hi-8	Filmed on BetaSP	Dubbed to one inch	Dubbed to BetaSP	Dubbed to Hi-8
Laserdisc	Filmed on BetaSP	Dubbed to one inch	Pressed laserdisc	

high-bandwidth analog formats, I believe that one extra generation had no impact on the ultimate test results.

We captured all footage with the new Intel Smart Video Recorder Pro, which has an updated analog front end featuring the new 7196 Philips chip. All video was captured at 15 frames per second at 320x240 resolution and compressed to 150 kB/second using Indeo 3.1.

Note that the Smart Video Recorder Pro has inputs for S-Video and composite, but not component. In fact, none of the capture boards that we tested offered component in.

As you would expect, both the BetaSP and the Hi-8 decks offered S-Video signals out. While the laserdisc stores the color information in composite format, the Pioneer CLD-V2600 also offers S-Video out. While this can't convert the composite color signal back to pristine S-Video, it allows the $1,000 deck to separate the colors rather than the $500 capture board. Accordingly, we used the Smart Video Recorder Pro's S-Video connector for all three sources.

The laserdisc puts out its own frame indicator, which is the number shown in the upper left-hand corner of the screen. Since we burned the frame numbers into the BetaSP tape after mastering the laserdisc, the time codes found in the other footage isn't there.

All files were captured at 320x240 at 15 frames per second. Capture formats were YUV-9 for the step captures and Indeo 3.2 for the real-time. All videos were filtered (see Chapter 11) and compressed to around 150 kB/second with the Indeo 3.1 Quick Compressor (see Chapter 12).

Here's what we focused on in comparing the results.

COLOR FIDELITY

We focused on two areas, the black coat and smooth gray back wall. For the coat, we looked for patches of green and other color errors. We judged the back wall based on the smoothness of the color and presence of streaks or other artifacts.

If you chose to follow along, we extracted frame 12:16 or 12:17 on the tape formats, and frame 13244 for the laserdisc. This is obvious in the first few frames, but not later.

COMPRESSION ARTIFACTS

The cleaner the signal, the fewer the artifacts. Zooming in on the face, we observed the smoothness of the facial features, especially around the eyes and mouth.

INDEO 3.1 QUALITY SETTING

We compressed all footage with the Indeo 3.1 Quick Compressor, which we'll introduce you to in Chapter 12. This codec is fast, works well in low-motion sequences and is still driven by the quality setting rather than the data rate. This means that you adjust quality to achieve your target data rate.

We always started at a quality setting of 100. If that didn't produce a 150 kB/second file, we dropped the quality setting to 90, and so on.

When forced to drop the quality setting, the codec is telling you "Hey, I can't work with this noisy video at the 100 quality setting and produce 150 kB/second—I'm gonna have to reduce quality." A video that compresses to 150 kB/second at a quality setting of 100 is therefore "cleaner" and less noisy than a video that requires a quality setting of 80.

That's the theory, anyway. In practice, we found that lower-quality formats required a quality setting of 80–90, while laserdisc and BetaSP worked fine at 100. Since the capture boards were the same, this indicates a stronger, cleaner signal coming from the two latter formats. For this reason, we'll report the quality settings of the videos as well as our subjective observations.

The Envelope, Please

Not surprisingly, the real-time BetaSP capture showed minor artifacting below the eyes and in other smooth regions in the face. This is the double compression at work (Fig. 7.7).

Overall, the laserdisc step frame was the champ (Fig. 7.8). The face shows good detail and smoothness, and the back wall is smooth with good color fidelity. The laserdisc showed little

Figure 7.7 BetaSP captured in real time (compression quality setting = 100)

Figure 7.8 Step capture from a laserdisc (compression quality setting = 100)

of the color instability I had feared would result from the composite signal. Conversely, the higher color quality of the BetaSP signal didn't translate into significantly higher-quality video. Perhaps this related to the S-Video signal in, perhaps to the double compression.

Both formats showed some color smearing, evidenced by pockets of green in the black coat. However, the detail on both videos was extremely crisp and sharp, even after compressing to 150 kB/second.

Third in overall quality was the Hi-8 step frame capture (Fig 7.9). While the single-frame picture quality was outstanding, we experienced several analog problems, such as "jitter," where the tape bounces up and down, and "drop-out," where spurious white lines randomly streak across the bottom of the picture. Both of these will be clearly evident in Fig7_9.avi.

During compression, jitter looks like motion to the codec, which retards compression. We had to turn the quality setting down to 75 to achieve the required data rate, which is the primary reason that Hi-8 step frame couldn't compete with laserdisc. In fact, the Hi-8 step-frame capture ended up looking worse than real-time Hi-8 capture because the real-time capture had no jitter and compressed more smoothly.

Figure 7.9 Hi-8 step frame capture (compression quality setting = 75)

These problems may be the delicate Hi-8 tape's not-so-polite way of telling us that it's not quite up to the rigors of step framing, even if the deck is frame-accurate and computer-controlled, thank you very much. On the other hand, I used this test as a learning experience and ran through it 15 or 20 times, which may be beyond the half-life of Hi-8 tape. The jitter and drop out seemed to turn up late in the game, and maybe wouldn't have appeared had I gotten it right the first time . . . or even the fifth time.

In addition, the test footage appeared on the first two minutes of the tape. We later learned that analog video experts advise not using the first five minutes of the Hi-8 tape for precisely these reasons. For your information, we captured the last five or six videos for the *Video Compression Guide and Toolkit* from the CVD-1000 in step-frame mode. The captured video was 30–40 minutes into the hour-long tape. Jitter was totally gone, and drop-out very infrequent, and the quality of the footage was close to, but not quite equal to, the laserdisc. To a degree, these subsequent results have convinced me to discount the test results presented here.

Overall, there's no doubt that laserdisc provides a physically stronger media that holds up better over multiple captures. On the other hand, read on, and you'll see that laserdiscs have their warts as well.

Format Hassles

You can shoot in Hi-8 format yourself, or shoot in BetaSP and
dub to Hi-8 for capture. Dubbing the BetaSP tape to Hi-8 can
be done in-house by most studios or post-production houses
and should cost under $50. In contrast, creating a one-write
laserdisc can be a real pain in the rear and costs between $250
and $500. You have to create a master tape that meets certain
specified requirements. Most analog production facilities don't
have a one-write laserdisc, so you may have to send the master
off to another city. The finished one-write disk only holds 30
minutes of video.

If you buy a *plastic* master ($250–300), the first three to
four minutes of video on the disk can be distorted. Color hue
and saturation tends to change as you move through the
laserdisc. *Glass* masters cost about $500, but offer better color
fidelity. How much better? Probably about 200 bucks' worth.
What's that mean? I don't know!

Notwithstanding these problems, many CD-ROM developers swear by laserdiscs and claim to have fewer problems than
I've had. If you decide to go this route, here's a list of steps to
take before cutting the master.

1. Advise the one-write facility in writing that you're using
   the disk for step-frame capture. There are two types of
   laserdiscs, CAV, for Constant Angular Velocity, and CLV,
   for Constant Linear Velocity. Most laserdisc players can
   only step-frame from a CAV disk, which has a capacity of
   30 minutes. You need to be certain that you're getting a
   CAV disk.

2. Get a written list of specifications for the master tape that
   you will provide to the one-write facility. This should
   include length of time required for tones and bars, which
   allow the one-write system to synch up the color signal,
   and length of time of pure black video. Keep this information on file and follow it to the letter.

3. Get a written description of the warranty for both the
   plastic and the glass masters. Determine the benefits of

the glass over the plastic and see what faults aren't warranted in the cheaper disk. The first few minutes of color on the first disk I cut was unstable. Later, when I complained, I was told that I had been advised of this up front. If you have less than thirty minutes of video to master, ask if you can place the video starting five or six minutes into the disk.

4. Don't purchase your master until all your footage is completed. Multiple laserdiscs are not only expensive, they also will usually look at least a little different from a color standpoint.

If you're working on a budget and doing your work in Hi-8 or SVHS, probably the best advice is to work in that format until all your video is finished and prototypes completed. Then, if you'd like to try and boost video quality, master the laserdisc and go from there.

To find the nearest facility, look up Video Tape Duplication and Transfer services in the Yellow Pages and call the company with the biggest advertisement. Normally, if they don't have a facility they will tell you who does. If you can't find a facility in town, try Crawford Post Production Services in Atlanta at (404) 876-7149.

Targa! Targa!

If step-frame is your game, sooner or later you'll start to wonder if a better capture card will improve video quality. No frame grabber has a better franchise than the Targa line from Truevision, so testing our Smart Video Recorder Pro against the Targa 64 was a natural. Here's what we found.

Once again, capture board differences make for color differences, and the Targa capture came out sort of overexposed. This relates more to capture settings than to capture board capabilities (mea culpa, mea culpa). Overall, however, the Targa board showed the best color fidelity of the bunch. The black coat is black, the back wall gray (Fig. 7.10).

Figure 7.10 Step framing with Truevision's Targa 64 (quality setting = 100)

Surprisingly, however, the actual analog front end responsible for capture seemed inferior to the Smart Video Recorder. Note the artifacts under the eyes and below the nose evident at normal resolution. At higher magnifications, the board exhibited streaking and loss of detail not evident in the Smart Video Recorder Pro.

The Targa 64 obviously has many features that the Smart Video Recorder Pro doesn't have. However, comparing apples to apples in the video capture arena, the Smart Video Recorder Pro seems to take the cake.

We'll discuss the mechanics of step-frame capture in Chapter 10.

THE GREAT ANALOG SOURCE TEST

Well, you have our advice—now let's see what it costs you *not* to follow it. If you're a real-time kind of guy or gal, can't wait in this step-frame stuff, don't want to cut a laserdisc, you'll probably want to know how the other formats stack up.

Working in IBM's analog multimedia facilities here in Atlanta, we created equal generation copies of BetaSP, S-VHS, VHS and Hi-8 tapes (see Table 7.3). The laserdisc was one generation younger than the other formats. Once again, all generations previous to the test tapes were high-bandwidth

Table 7.3 Generations of test formats

Tested Format	1st Generation	2nd Generation	3rd Generation	4th Generation
BetaSP	Filmed on BetaSP	Dubbed to one inch	Dubbed to BetaSP	Dubbed to BetaSP
Hi-8	Filmed on BetaSP	Dubbed to one inch	Dubbed to BetaSP	Dubbed to Hi-8
Laserdisc	Filmed on BetaSP	Dubbed to one inch	Pressed laserdisc	
S-VHS	Filmed on BetaSP	Dubbed to one inch	Dubbed to BetaSP	Dubbed to S-VHS
VHS	Filmed on BetaSP	Dubbed to one inch	Dubbed to BetaSP	Dubbed to VHS

analog formats and I believe the multiple generations had no impact on the final results.

We tested with three different capture boards to normalize our results. These were the VideoLogic DV4000, the MiroVIDEO DC1 tv and Intel's Smart Video Recorder Pro. Here are our observations.

Figure 7.11 shows clipped video segments zoomed to 200 percent in the respective formats. Overall, the clearest, smoothest image was produced by the laserdisc, even though we couldn't step-frame capture with the DV4000 because its capture application didn't provide step-frame control. The best of

Figure 7.11 VideoLogic DVA4000 working with the respective formats

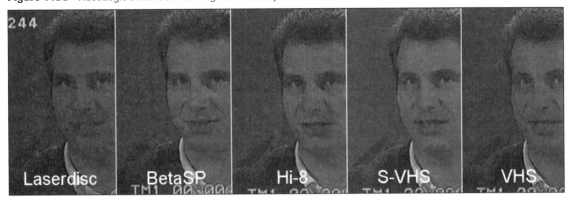

Table 7.4 Compressor quality settings for VideoLogic DVA4000 in respective formats

Format	Laser	BetaSP	Hi-8	S-VHS	VHS
Compression Quality Setting	100	100	90	90	90

the rest was clearly BetaSP, which showed fewer artifacts and less streaking than any other format.

Streaks on the back wall become evident with the Hi-8 video, which appears to be of slightly higher quality than the S-VHS in this instance. In the original videos, S-VHS showed fewer artifacts than Hi-8. However, the S-VHS showed less blurring around the chin and looks better in the picture in Fig. 7.10. Overall, VHS was clearly the worst-looking image, with blotches of green in the black coat, a totally mottled back wall and fairly apparent facial artifacts.

The respective videos are included in the Chapter 7 subdirectory of the CD-ROM as DVABETA.AVI, DVAHI_8.AVI, DVASVHS.AVI, DVALASR.AVI and DVAVHS.AVI.

MiroVIDEO DC1 tv

The Miro board help up surprisingly well through the first three formats with only minor differences between BetaSP, Hi-8 and S-VHS. Quality for the VHS format is definitely a step below that of the other three.

In the great battle between the medians, I pick Hi-8 over S-VHS both in Fig. 7.12 and in the original videos. Probably wouldn't throw away my S-VHS deck and buy a new Hi-8 player, however, because the quality difference is minimal (Table 7.5).

The respective videos are included in the Chapter 7 subdirectory of the CD-ROM as MIROBETA.AVI, MIROHI_8.AVI, MIROSVHS.AVI and MIROVHS.AVI.

The Intel Smart Video Recorder Pro

No surprises here (Fig 7.13). The laserdisc again produced the cleanest back wall and most artifact free finish. The black jack-

Figure 7.12 Analog formats captured with MiroVIDEO DC1 tv

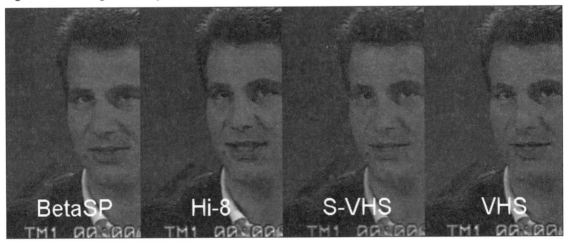

Table 7.5 Compressor quality settings for MiroVIDEO DC1 tv in respective formats

Format	BetaSP	Hi-8	S-VHS	VHS
Compression Quality Setting	100	100	100	100

et was black through and through. The BetaSP also showed excellent color fidelity in the jacket and wall.

S-VHS shows fewer artifacts than Hi-8, but greater irregularity in the back wall. S-VHS bled more green into the black coat than Hi-8. VHS clearly lags all the formats, showing gross

Figure 7.13 The Intel Smart Video Recorder Pro runs the analog format gauntlet

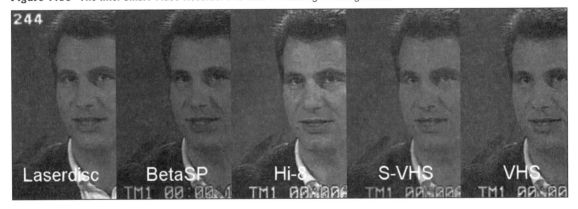

Table 7.6 Compressor quality settings for Intel Smart Video Recorder Pro in respective formats

Format	Laser	BetaSP	Hi-8	S-VHS	VHS
Compression Quality Setting	100	100	90	90	80

break-up and artifacting in the face and streaking in the back wall (see Table 7.6). The respective videos are included in the chap_7 subdirectory of the CD-ROM as ISORBETA.AVI, ISORHI_8.AVI, ISORLASR.AVI, ISORSVHS.AVI and ISORVHS.AVI.

Summary

What have we learned? Well, I recall a politically incorrect, not funny then, not funny now, but very appropriate kind of joke. It goes like this.

What's a six, then a seven, then an eight, then a nine and finally a 10? The answer? A person of the opposite sex in a bar on Saturday night at 9:00 PM, 10:00 PM, 11:00 PM, 12:00 AM and finally at 2:00 AM.

In the digital video world, it's the same question, but the answer is—your video on VHS, S-VHS, Hi-8, BetaSP and laserdisc. Clearly your results will suffer if you capture from VHS, especially if you film in that format as well.

If you work in BetaSP, S-VHS and Hi-8, you can improve your video by transferring the video to a laserdisc and step-capturing. Nothing against these formats, it's simply the difference between double compression and single compression.

My company will probably capture from Hi-8 from now on because the ultimate difference between BetaSP and Hi-8 is minimal, and the more time I spend with the CVD-1000 the more I think that Hi-8 step-frame capture is for real. When push comes to shove, however, and video quality really counts, I just might sweat through another laserdisc mastering.

What else did we learn? A gray flat background is probably the worst background for filming for video compression. Gray is obviously not a primary color, and tends to absorb color

Figure 7.14

from surrounding objects. In the various captures, sometimes it turned green, sometimes red.

It also seems as if codecs abhor large smooth areas like the background in these videos, and attempt to fill them with detail that isn't there. It's better to use a busier background like that presented in Figure 7.14, which is filled with objects colored mostly brown and other earth tones. Stay away from highly detailed backgrounds, which also tend to create motion.

Next chapter we continue our quest for the perfect capture when we analyze capture cards, capture stations and other alternatives to video capture.

SUMMARY

1. VHS, BetaSP, Hi-8, S-VHS and laserdiscs are analog filming and storage formats with different characteristics relating to color information storage, lines of resolution and signal strength. These characteristics translate directly to capture and compression performance.

2. There are three types of video signals. Component video signals maintain three separate channels for color information, which reduces video noise. The only reasonably

accessible component filming format is BetaSP. S-Video signals maintain two channels, a chroma or color channel and a luminance channel for black-and-white information. This format, available on S-VHS and Hi-8 formats, is cleaner than composite, which maintains one channel for all video information. Composite formats include VHS, 8 mm and laserdisc.

3. Signal to noise ratio quantifies the strength behind the signal. If you anticipate multiple video generations, you should film with a stronger format such as BetaSP, or transfer lower formats to BetaSP before editing.

4. The quality of your camera and lens translates to ultimate compressed video quality. Prototype with inexpensive cameras, but shoot your final footage with as high-quality a camera as possible.

5. Step-frame capture avoids compression during capture. To step-frame from an analog source, it must be frame-accurate and computer-controlled through a protocol supported by your capture software. The two most accessible step-frame devices are most laserdiscs and the Sony CVD-1000 Hi-8 deck.

   Our tests revealed that step framing from a laserdisc produced superior results to real-time capture from BetaSP, proving that signal strength and color fidelity don't outweigh the negatives of double compression. If you decide to go the laserdisc route, take the following steps:

   (a) Get a CAV disk.

   (b) Request written specifications for the master analog tape from which the laserdisc will be cut. Follow them to the letter. Request a written explanation as to the advantages and disadvantages of glass vs. plastic disks.

   (c) Understand your warranty.

(d) Don't purchase a master until you're sure that filming is complete.

6. Results from the CVD-1000 showed jitter and drop-out. However, single-frame video quality was excellent, and subsequent tests revealed that the analog faults could have related to overstressing the tape or to capturing too close to the start of the tape.

   Other multimedia producers have also reported good success with this technique. Hi-8 format is much more accessible than laserdisc, and you should at least experiment with this technique. Make sure you start all filming at least 5 minutes into the Hi-8 tape.

7. In step-frame tests, a Targa 64 frame grabber produced better color fidelity but more artifacts than the Smart Video Recorder Pro. Purchasing the more expensive card for step-frame capture will not produce overall higher-quality video.

8. In real-time tests performed with three different capture boards, the BetaSP format was always superior to the other three tape-based formats. Hi-8 seemed to have a slight edge over S-VHS in most tests, and VHS always produced the worst results.

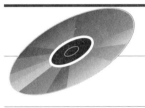

CAPTURE KARMA II

CAPTURE KARMA II

This is the capture board chapter. We start by analyzing features to look for in your capture board. Then we'll take a look at several mid- and high-end ISA-based capture boards to see how they compare.

At the next level are EISA boards, and we'll look at output from the Targa2000, Truevision's most recent product entry. We'll also examine what you get when you cross a Smart Video Recorder Pro with 64 Megabytes of RAM.

Finally, the holy grail. The highest quality captured footage we could beg, borrow and steal; and we did a lot of the first two, and a little of the third to assemble this collection of captured video.

ANATOMY OF THE PERFECT CAPTURE CARD

Last chapter we learned how to avoid compressing during capture. But if you can't like step capture you've got to compress on the way in. So it's time to pick a capture board.

We looked at seven or eight boards over the course of a two-month period. The four we decided to review are, the Fast Movie Machine Pro with M-JPEG Option, Intel's Smart Video Recorder and Smart Video Recorder Pro, Miro's MiroVIDEO DC1 tv and VideoLogic's DV4000. The cards ranged in price from under $649 to around $3000.

The boards are all ISA-based boards since these are most common. We'll look at the results of an EISA-based board, the Targa 2000 at the end of the chapter. Then we'll look at two other capture options including the Holy Grail of digital video.

Here's a summary of our likes and dislikes that provides a checklist of features to help you analyze future purchases. We'll evaluate each board separately later in the chapter.

On-Board Hardware Compression

The first job of the capture board is to get the frames on disk. This means capturing your target data rate at your target resolution without dropping frames.

The early flood of cheap capture boards over-promised and under-delivered. Most promised 30 frame per second capture at 320x240, and very few could even provide 15 fps. That's because most early contenders had no on-board compression. As we've seen, the ISA bus can't transfer and store anywhere close to the 3.5 MB/second bandwidth of 15 fps 320x240x24-bit video. Unless the board could compress the video 10:1 on-the-fly, it would inevitably drop frames. And drop frames the early contenders did.

Introduced in the summer of 1993, Intel's Smart Video Recorder set the bar for all future capture boards, primarily because of its I-750-based on-the-fly Indeo compression. This on-board hardware enabled the Smart Video Recorder to capture 15 fps at 320x240 without dropping frames.

With the new Smart Video Recorder priced at $570, there's really no excuse not to purchase a board with on-board hardware compression, even if you plan to step-frame capture. It's simply too valuable a feature for prototyping and other quick-turnaround captures.

All tested boards provide hardware compression. The two Smart Video Recorders featured Indeo 3.2, the others Motion JPEG. We tested hardware performance two ways. First, to measure pure ability to get frames to disk, we captured at 15 and 30 fps. Then we measured capture quality by comparing

15 frame-per-second videos captured from the BetaSP deck and compressed to identical parameters.

Bus

Let's stay with 15 frames per second at 320x240 resolution for a moment. As we'll see later in this chapter, an EISA or PCI bus board captures higher-quality video because it only has to compress the 3.5 MB/second stream down to about 2 MB/second, less than 2:1. Thus, a PCI-based capture board can deliver near step-frame quality in real time. In contrast, an ISA capture card must compress the video down to between 350 and 700 KB/second, or in the range 5–10:1.

As of August of 1994, there was only one announced PCI capture board, the Matrox Marvel II, which costs $895. It was not available for testing in time to include in this book. However Matrox builds great products, and their current very high-end capture cards are among the best on the market.

By early 1995, however, most serious players will have announced and some will have shipped PCI-based cards. The difference in potential capture quality should warrant a close look at these new cards, especially those from reputable players such as Matrox. Besides, it will give you an excuse to purchase that PCI computer you've been wanting.

S-Video Input

You would think that boards costing over a thousand dollars would be guaranteed to have S-Video in as well as composite. But you'd be wrong.

The Fast Movie Machine Pro, which was designed primarily to drive an analog tape editing system, only has composite in. Similarly, the DVA4000, an overlay card whose main purpose in life is to display analog signals on a digital computer screen, offers composite as the standard input signal—you have to request S-Video.

Installation Software

OK. There is software that scans for open interrupts, DMA channels and I/O ports. There are capture boards that require these links into your computer memory and processor. These boards have installation software.

So. Why doesn't the installation software scan for open interrupts, DMA channels and I/O ports? Don't know, but they all sure should, and most don't.

Figure 8.1 shows the installation screen from the Miro-VIDEO DC1 tv. Not only did you have to select an I/O port, interrupt and DMA channel, you had to exclude upper memory from EMM.386 and in WIN.INI.

When I was at IBM testing these facilities, an IBM engineer and a buddy who's also an engineer helped with the installation and testing of the boards. It took three of us over two hours to install this board. After trying about 15 video memory ranges with no success, we ended up uninstalling EMM.386. This hosed my SoundBlaster AWE32 installation,

Figure 8.1 The installation gauntlet of the MiroVIDEO DC1 tv

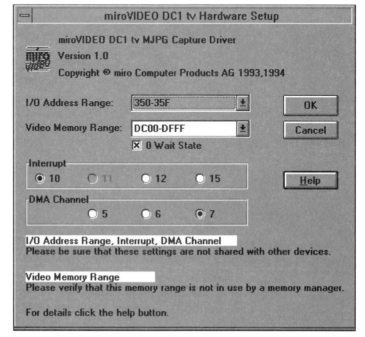

so we had to reinstall my old 8-bit SoundBlaster. Later, by calling tech support, we got our board installed, but had to change our system bios to do it.

To add insult to injury, a few days later I read a review of the board in a prominent video magazine which said, "Installing the board is literally a snap."

Literally? Literally? I suppose the reviewer must not have meant Webster's first definition of snap, which is "to make a sudden closing of the jaws." No mentions of wounds or close encounters. Or the second, which is "to grasp at something eagerly." Since the board technically is a frame grabber, maybe this was the one.

No, I suppose he meant the 25th definition, "something that is easy and presents no problem." If so, however, I'm not sure how "literally was a snap" is different than "was a snap." But leaving this petty frustration aside, what kind of machine was this installed in? I mean, this board has more lines in and out than a patient in intensive care.

Let's look at DMA channels. You start life with eight. Your floppy drive takes one, your hard drive another, one cascades (whatever that is). The SoundBlaster AWE32 takes two more. So you're left with better than a sixty percent chance of a conflict.

Maybe the other guy is living right and installed the board in two computers without a DMA conflict. Maybe my computer has peripheral gremlins that conspire to create conflicts. Either way, there's no excuse for an installation this difficult without an installation program to help you along the way.

The Smart Video Recorder Pro was the only board we tested that searched for open slots. The Fast Movie Machine Pro helped by allowing you to test the interrupt and I/O port settings during the installation, which is almost as nice.

Hardware Diagnostics

When I first installed the Smart Video Recorder Pro, I had trouble getting video to show up in VidCap. So we ran Intel's diagnostic program, shown in Fig. 8.2, and traced the problem to a faulty S-Video cable. By localizing the problem, the diagnostic routine saved hours of debugging and testing. Intel was

Figure 8.2 Intel Smart Video Recorder Pro's helpful diagnostic program

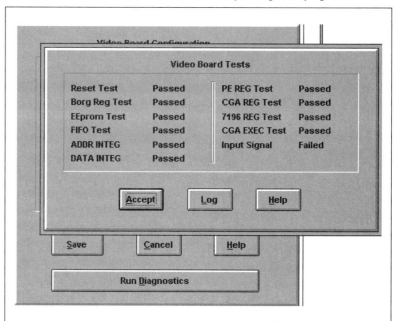

the only vendor offering a formal diagnostic program. Video-Logic's DV4000 offered some diagnostics, but they were not as comprehensive or as accessible.

Uninstall Program

Uninstall programs uninstall the drivers, Windows groups, entries in your WIN.INI file and other fallout of the capture board's existence in your computer. Of all the boards that we tested, only the Smart Video Recorder Pro had an uninstall program.

I don't know—maybe board vendors think that an uninstall program is bad luck, kind of like a prenuptial agreement. But you'll end up dating more capture boards than you marry, and you'll want to escape with as few random DRV and DLL files on your hard drive as possible. Without an uninstall program, not even a lawyer can help you.

At last count, our main capture station has 96 DRV files and 156 DLLs, and I'm certain that most aren't doing anything but taking up space. I hope that number is limited by future capture boards, graphics boards and other products that come with uninstall programs.

Capture Formats

The primary benefit from step framing is avoiding double compression, so you'll want your capture board to be able to capture in an uncompressed format. While most boards with onboard compression offer compressed and uncompressed capture options, some don't, most notably the Fast Movie Machine Pro. In fact, with the Movie Machine Pro you have to compress your incoming video at least 16:1, far more than is necessary to get 15 fps 320x240 video on your disk without dropping frames.

At the very least, your capture board should offer a YUV color reduced format, and most do. YUV format is usually between 30% and 60% more compact than uncompressed RGB. Speaking of RGB, it's also nice to be able to capture in this format, since it lets you verify that your YUV format is accurate.

For example, we have worked extensively with VideoLogic's Captivator, not reviewed here because it doesn't offer hardware compression. The YUV format on the Captivator is greenish in appearance, and the difference between RGB and YUV is very apparent. Accordingly, we always captured in RGB. The Smart Video Recorder offers only YUV and compressed Indeo 3.2—no RGB (see Fig. 8.3). I prefer to capture in both YUV and RGB and make my own decision.

Incoming Color Control

As we saw last chapter, analog color varies widely between formats and even decks. Incoming color control lets you modi-

Figure 8.3 The Smart Video Recorder Pro offers only two capture formats, Raw, which is actually YUV, and Indeo 3.2. I would prefer RGB as another option

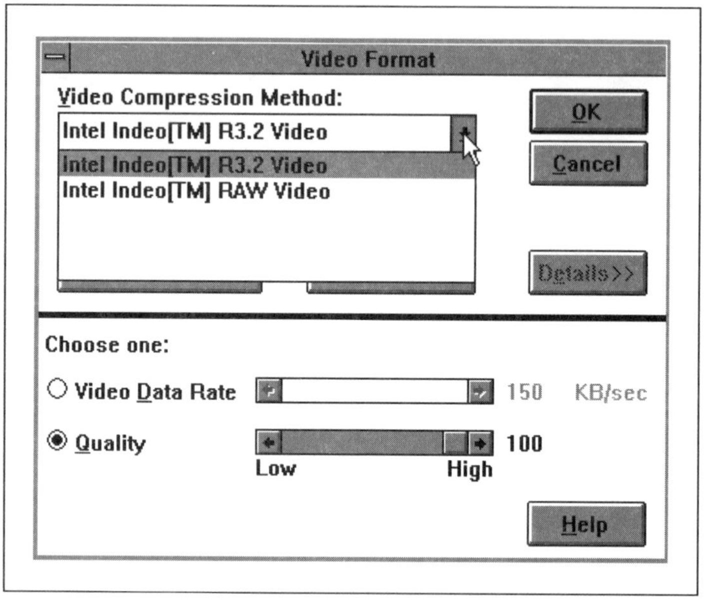

fy the analog signal before the video hits your hard drive, which makes for better-looking video.

Color controls typically come in two flavors, those like VideoLogic's that offer control over Saturation and Hue, and those that offer direct manipulation of the red, green and blue signals (see Fig. 8.4). I find the former easier to work with in most instances. Brightness and contrast are usually offered as separate options on all systems that offer color balancing, whichever style.

At this point, you probably shouldn't purchase a product that doesn't offer color control. The incoming signals are too variable, even over the life of a single tape or laserdisk. If you can't adjust the signal on the way in, you'll be forced to adjust the video in Adobe Premiere, which is both difficult and time-consuming.

All boards offer some form of color and brightness adjustment except for the Fast Movie Machine Pro, and it's a glaring omission. What's particularly frustrating is that Fast allows

Figure 8.4 VideoLogic's color control

you to control the video overlay signal, or the video that shows up on your screen, but not the captured digital signal. So you end up spending valuable time adjusting the incoming video signal to just the right parameters, only to learn through a call to technical support that it's all for naught.

Multiple Input Resolutions

The capture board should allow multiple input resolutions in all formats. For example, the old Smart Video Recorder didn't offer 240x180 resolution in raw format. If you wanted to step-frame capture in that resolution, you were out of luck. Since the *Video Compression Guide and Toolkit* uses 240x180 resolution video, we were out of luck.

Fortunately, Intel has remedied this in subsequent releases, but remains fairly rigid in capture resolution. Other boards are more flexible. As shown in Fig. 8.5, you only have three options when capturing in compressed Indeo 3.2 mode, and raw mode offers only one additional resolution, 640x480.

While you can remedy this to a certain degree by cropping in VidEdit, it's an unnecessary step. If you plan on working

Figure 8.5 The Smart Video Recorder Pro's limited input resolution choices

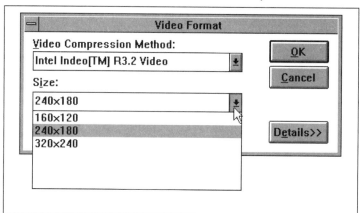

with irregular capture formats, Intel's lack of flexibility remains a drawback.

Supported Broadcast Standards

Make sure your board supports your local broadcast standard, whether it be NTSC, PAL or SECAM. For the overwhelming majority of readers, multiple standard support won't be an issue. However, virtually all the boards support both NTSC and PAL, the two primary standards. If you need SECAM, be sure that it's supported before purchasing.

Bundled Software

The software bundled with the board also bears consideration. You'll need software to control the capture operation and perform rudimentary editing. Most boards are compatible with VidCap and VidEdit, but may not come bundled with these programs.

If the board doesn't include Video VidCap and VidEdit for Windows, it will typically bundle either Adobe Premiere, which includes a capture application, or another video editor that also includes a capture application. The more retail-oriented products, such as Intel's, also tend to bundle a multimedia or presentation program like Asymetrix Compel. While only somewhat tangential related to the capture and compression function, Compel is a great tool and a nice addition to any capture board.

Company History

While not really an issue with any of the boards reviewed here, it pays to buy from a reputable company. Capture board vendors have sprung up like microwave popcorn, and some disappear as quickly. MediaVision's Pro Movie Spectrum won a *PC Magazine* Editor's Choice in April 1994 and was withdrawn from the market in June.

Drivers and compatibility are a major issue with all multimedia peripherals. If you're purchasing an ISA bus product, you'll probably move to PCI in the next 12 to 18 months, which lessens the risk of your vendor going away. However, if you select a PCI board, you'll want to keep it for three or four years. This means that you'll need three or four driver revisions to stay current with new operating systems such as Chicago, Windows NT and future codec upgrades.

Support is another issue—these products don't install themselves, and problems may arise long after date of purchase. Intel has a strategic interest in Indeo's success and will be around to support their customers. Most other vendors are in the business strictly for commercial reasons.

So resist the urge to purchase hot hardware from a slick newcomer. In the long run, it may pay to stay with old reliable.

Other Features

There's a host of other features of potential interest to video producers, including video overlay capabilities, where the board can circumvent the computer's digital graphics system and place analog video directly on screen; video encoding capabilities, which allow the board to lay video or animation to an external VCR or other device; and the ability of the board to play back captured video in real time and serve as a video server for a kiosk or other similar applications.

However, in our relentless pursuit of the ultimate capture system, we primarily focused on issues relating to ease of use and pure video capture capability.

BOARD REVIEWS

How did the boards stack up? Here's a thumbnail sketch of the boards we actually looked at in-house, covering the Motion JPEG boards first, then the Intel Smart Video Recorder father and son combination, and then the Captivator. As mentioned earlier, we tested the board's pure capture ability by capturing at 15 and 30 fps. Then we compressed the 15 fps file to compare the quality of the captured video.

Fast Movie Machine Pro

Based in Germany, Fast Electronic GmbH has almost a mythical reputation among analog video producers for their high-quality video editing systems. The Movie Machine Pro is no exception, and it is a superb device for video overlay and video editing—but not, unfortunately for video capture.

The Movie Machine Pro can display an analog signal on a computer screen, either from its own integrated television tuner or from any other analog source. In addition to watching your favorite program from your desk, you can also grab screen bit-maps for later use.

The board's primary role in life, however, is to edit analog movies. The Movie Machine software supplies transition effects, and a library of special effects, and it enables titling. With an external destination VCR, you can take in analog video from the TV tuner or another VCR, edit it, and then record on the second VCR.

However, the Motion JPEG Option which we used for video capture seems almost like an afterthought—as if the engineers were sitting around thinking, "Jeez, we've got this great board that works with analog video, why don't we add a Motion JPEG capture card?" And limitations in the main board that don't affect editing performance doom the Motion JPEG capture option to second-rate performance.

Specifically, the Movie Machine Pro only accepts composite signals in—not a great loss if your primary goal is laying editing analog video to tape. However, as we've seen, a composite signal is noisier than an S-Video signal. During our testing, we captured composite and S-Video with the boards that had both connectors, and the difference was apparent—S-Video produces cleaner, clearer video.

As previously mentioned, the Movie Machine Pro also forced you to compress the incoming video by at least 16:1 during capture. Consequently, our capture bandwidth at 15 frames per second was 265 kB/second. At 30 frames per second, bandwidth was 487 kB/second, illustrating the capacity of the capture station to handle much more than the 265 kB/second allowed by the capture hardware. Had the board captured at 15 fps at 487 kB/second, which the capture station was capable of doing, video quality would have been much higher.

How did this affect quality? Well, the Movie Machine Pro was the only Motion JPEG board to exhibit gross evidence of the so-called "Gibbs Effect," a JPEG artifact characterized as distortion around numbers or letters (see Fig. 8.6). This isn't surprising, considering that the other two Motion JPEG boards captured at least 50 kB/s more data per second. FAST's design decision clearly cost it in quality.

The Movie Machine Pro also didn't let you adjust the color or brightness of the incoming video signal. When we called FAST technical support, they recommended that we use the included Adobe Premiere perform this operation. While Premiere is a nice feature, the best solution would be to allow editing of the incoming signal and avoid the post-processing.

Figure 8.6 The three Motion JPEG boards. Only FAST, which forces dramatic compression upon capture, produces gross evidence of the Gibbs effect, a JPEG artifact

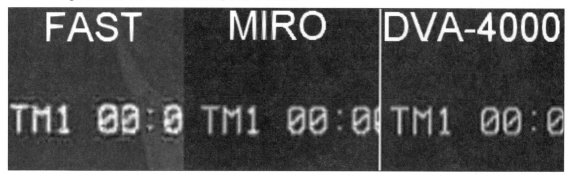

In terms of pure video quality, FAST lagged behind the other two Motion JPEG contenders (Fig. 8.7). It was also the only capture board of the three that required a compression quality setting of 90 to achieve the 150 kB/second data rate.

From a pure performance standpoint, the FAST board performed well, getting 30 frames per second to disk without problem. We also liked its installation program—while it didn't go out and find interrupts and DMA channels for you, it let you test them in real time, which is almost as good. The FAST software also tests the throughput capacity of your system, another nice feature.

On the negative side, the board doesn't offer a raw capture format—if you're a step framer, you're limited to the same 16:1 compression. The software also had an annoying habit of initializing configured for PAL rather than NTSC—understandable, given the European origin of the product. However, when you started playing video with the PAL configuration, the software ended up crashing with the cryptic error message

```
Target code 19008B> 16384B changed to 16384B
```

and locking the computer. We ended up reinstalling Video for Windows, then Windows itself to try and fix the problem, then discovered that the simple solution was checking "NTSC" instead of "PAL" whenever you loaded the capture application.

Figure 8.7 Video quality for three motion JPEG boards

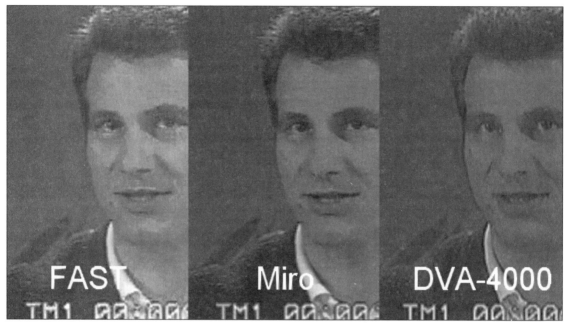

The Movie Machine Pro is a wonderful analog video editing system. However, as a capture station, the boards simply doesn't measure up. The FAST BetaSP capture file is in the Chapter 8 subdirectory under FASTBTSP.AVI. The Movie Machine Pro with Motion MPEG option costs $940. Contact Fast Electronics at (415) 802–0772.

MiroVIDEO DC1 tv

I've already described the problems we had installing the MiroVIDEO DC1 tv, another product developed in Germany, and I won't repeat them. I will say that when we finally got the board working, the thought that immediately leaped into my mind was "kidney stone." Yes, kidney stone.

Here's the story. About five years ago, in a previous life, I was consulting with a small company who had contracted long-distance services from a telephone company here in Atlanta. Service was poor because of delays in the installation of a telephone switch in Denver.

On the 17th floor of a new glass and steel building overlooking Lenox Mall, in a beautiful conference room by a rosewood table, the natty phone company president told us that installing a switch reminded him of passing a kidney stone. He described the worst pain ever experienced, and recollected that he had spent three days jumping up and down, screaming and crying, on the floor, on the table, in the halls, in the elevator, both in pain and in the hope of helping gravity move the stone through his internal plumbing.

Then, the stone passed, and it was like it had never happened. The world was a beautiful place, life was wonderful, and that's exactly how it was going to be after the switch came up in Denver.

Well, installing the MiroVIDEO DC1 tv was like passing a kidney stone. Once we got it installed, the world was a beautiful place and everything was wonderful.

The DC1 neatly sidestepped the potholes that hindered the FAST Electronics board. It accepted both S-Video and composite input. It captured in both RGB format for step-frame capture and Motion JPEG. It allowed you to modify the incoming signal for color and brightness. It included a utility that measured system throughput but didn't force you to work with predetermined capture limits.

The DC1 was the only one-slot Motion JPEG capture solution, evidencing its design as an integrated capture solution, not an add-on for another video product. However, its video overlay capabilities were on par with FAST's and almost equal to those of the VideoLogic board.

From a pure performance standpoint, the DC1 captured both 15 and 30 frames per second without problems at file bandwidths of 308 and 525 kB/second, respectively. The latter was the highest bandwidth we achieved on the capture system.

From a quality standpoint, the DC1 was the only board that captured all four tape formats and compressed them to the target data rate at an Indeo quality setting of 100. Not only do the videos look subjectively better than the other boards, the Indeo codec thought the signals were cleaner as well.

My only complaint with the DC1 was a minor one—we had to adjust the color and brightness settings several times to get the video looking right. This, however is like complaining about rain in Seattle: It's just part of the game.

Bundled software includes Video Studio from U-Lead Systems, the original developer of Aldus Photostyler. Video Studio includes video capture and editing capabilities, neither of which I tried since the DC1 works with VidCap as well. For an additional $50, you can also purchase Adobe Premiere. Also bundled with the board is a 24-bit image editing program called iPhoto Plus.

All this was almost enough to make you forget that you had just passed a . . . er, that the installation was extremely cumbersome. And the fact that we had to change our system BIOS to make it run at all. Overall, the MiroVIDEO DC1 topped our charts for video quality and pure performance. The MiroVIDEO DC1 tv retails for $849 and is available directly from the company at (415) 855–0940. The Miro BetaSP capture file is in the Chapter 8 subdirectory under MIROBTSP.AVI.

VideoLogic DVA4000

The primary function of VideoLogic's DVA-4000 is video overlay, or the placement of analog video on a digital video screen. As one of the first such boards, the product has a long and truly venerable history and is installed in thousands of kiosks, training and other similar applications around the world. First introduced by the English company in 1990, the DVA-4000 was multimedia before there was multimedia.

Judged as a pure overlay card, the DVA-4000 offers unparalleled performance, with clear analog video up to full screen. However, while it juggles the capture task more adroitly than the FAST board, pure capture performance falls behind that of other products reviewed here. While it makes sense to consider the capture option if you're already working with the DVA-4000, it probably doesn't make sense to purchase this expensive board set strictly to capture video.

Mechanically, the DVA-4000 was the most challenging product to install. VideoLogic recommends swapping out your video card, and supplied a Diamond SpeedStar that worked quite well. There are two other cards to install, plus an assortment of terminators, connectors and cables sufficient to intimidate the faint of heart. However, it all fit together remarkably easily, and we were soon rewarded with better than television-quality video displaying on our computer screen.

VideoLogic offers two cabling schemes, one for S-Video and one for composite. We tested with the S-Video option.

VideoLogic's software measures hard drive capacity and limits capture throughput. Notwithstanding these limits, the DVA-4000 was the only Motion JPEG solution that dropped frames at 15 fps capture. If you flip through the DVA4BTSP.AVI on the CD-ROM, you'll see some frames repeated and some irregular intervals where the board jumped to the third frame out instead of the second.

Worse yet, it didn't tell us it was dropping frames, which would have let us adjust input parameters and perhaps cure the problem. We didn't notice the dropped frames until well into post-production.

The captured video itself was slightly gauzy—kind of like those slow close-ups of female leads in old black-and-white movies. It compressed to a subjective quality level just below that of Miro. The video compressed to 150 kB/second at an Indeo quality setting of 100, matching Miro's performance.

VideoLogic offers a most comprehensive suite of proprietary software which we found very easy to use. The software excels in capture setup, allowing full control over input parameters, screen placement and capture formats (raw and compressed). However, it didn't support step capture and the board wasn't designed to work with VidCap. Diagnostic software was a bit more obtuse, but helpful nonetheless.

Overall, working with the DVA-4000 was like riding in a Rolls-Royce or old Jaguar. While it's a truly beautiful ride and you're always comfortable and well appointed, you never really trust the hardware. As a video overlay card, the DVA-4000 is still a great product. From a pure capture perspective, it's getting long in the tooth, and needs a major hardware overhaul to compete with solutions costing thousands less.

The DVA-4000 costs $2,995. Contact VideoLogic in Cambridge at (617) 494–0530. The VideoLogic BetaSP capture file is in the Chapter 8 subdirectory under DVA4BTSP.AVI.

The Smart Video Recorder Pro

At an estimated street price of under $500, there was never a question that the Smart Video Recorder Pro would be the low-end market leader. The only questions were, did it pay to buy the new board if you already owned the old, and how much did you have to spend to capture at a higher video quality?

The answers are yes, and $849. If you own the old Smart Video Recorder, you probably should think strongly about picking up the new product, especially if Intel offers any attractive trade-ins. But while the Intel Smart Video Recorder Pro performed admirably in our tests, it just missed topping the chart to the Motion JPEG-based MiroVIDEO DC1 tv. If, of course, you can survive the installation.

The Smart Video Recorder Pro is a cost-reduced version of the old board featuring a new Philips front-end chip and a clock doubled I-750 real-time Indeo compression chip. The old board was an extremely busy motherboard/daughtercard set, while the new board is a clean, one-board design that just looks tons cheaper to make. Thus the retail price reduction from $699 to $570 and an expected street price of under $500.

The Smart Video Recorder Pro uses the Philips Video Decoder 7196, the same "front end" as the Miro board. This is the chip that performs most of the analog-to-digital conversion. During development, Intel went to great lengths to achieve the optimal color and brightness settings for the chip, and it shows—while Intel lets you modify the incoming video signal, usually you don't have to.

The board features both S-Video and composite connectors. As we discussed earlier, the installation program helps you find open interrupts and I/O ports, but there are no guarantees the board will run if installed in open ports. The first time we installed the board we configured the software to the same parameters as the old Smart Video Recorder and the board didn't run. We ran out of time and moved on to another board. Later, we changed to a different setting which worked flawlessly.

The board ships with the Digital Video Producer, an upgraded version of the Splice video capture and editing program, which was recently purchased by Asymetrix and given a face lift. We used DVP extensively in Chapter 11 and found it easy to learn and extremely functional. The Pro was the only board to ship with an uninstall program.

Intel claims the new board will capture 30 frames per second at 320x240 resolution, which we couldn't duplicate on our test computer. However, we tested an early beta unit—perhaps the final version will boost performance to that level. The board captured 15 frames per second at 320x240 without problem at a bandwidth of 391 kB/second. Since you won't typically work with frame rates in excess of 15, failure to capture at 30 frames per second should not be a problem.

From a quality standpoint, the new board is a clear improvement over the old in both real-time and step-frame mode. In real-time performance, shown in Fig. 8.8, the Pro produced clearer time code lettering, and fewer artifacts above the lips and under the eyes. If you examine the two videos side-by-side, you'll also see that the background is more consistent on the Pro, albeit somewhat darker. Finally, even Indeo thought the new board produced a clearer capture signal—the video produced by the ISVR Pro compressed to 150 kB/second at a quality setting of 100, while the old board required a reduction in quality down to 90. (Video files are PROBTSP.AVI and OLDBTSP.AVI.)

In step-frame performance, the quality of the new Philips chip really shows (Fig. 8.9). The Pro produces a much clearer, more consistent background than the older model and is almost completely artifact-free. The colors bleeding into the coat on the older model are also no longer in evidence. Finally, the Indeo compression quality setting on the Pro was 100, while we had to back down to 90 to compress the video captured with the old board down to 150 kB/second. (Video files are PROSTEP.AVI and OLDSTEP.AVI.)

Clearly, Intel improved capture performance on the new board, and if the older model is your current capture board,

Figure 8.8 The battle of the Smart Video Recorder in real-time capture. Check out the numbers at the bottom of the page, and the area above the lips and under the eyes. The Pro shows fewer artifacts and an overall clearer picture

Figure 8.9 Old vs. new in the step frame battlefield

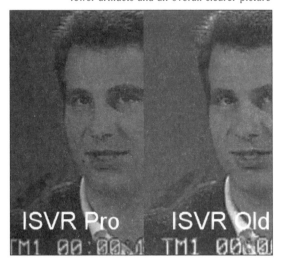

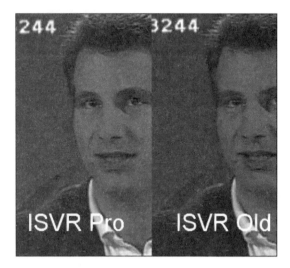

it's time to think about upgrading. The question is, do you spend $500 and get the new model, or $850 and get the MiroVIDEO DC1 tv?

MiroVIDEO DC1 tv vs. Smart Video Recorder Pro

From a pure quality standpoint, you get the Miro board. From BetaSP (Fig. 8.10) to VHS (Fig. 8.11) the Miro board produces higher-quality compressed images, not only visually, but objectively—the VHS file captured by the Smart Video Recorder Pro had to be compressed at a quality setting of 80 to meet the target data rate. In contrast, the VHS file captured by the MiroVIDEO board compressed at a quality setting of 100.

The Smart Video Recorder Pro will be about $350 cheaper than the Miro board. It should also be simpler to install and

Figure 8.10 Miro VS ISVR Pro-BetaSP

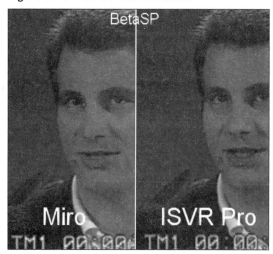

Figure 8.11 Miro vs. ISVR Pro—VHS

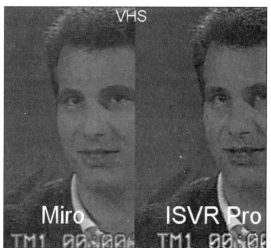

support and should fit into your current environment much more smoothly, which counts a lot when you're working with a board day after day. Intel has also been very forthcoming with new upgrades and enhancements, and plans a major upgrade to Indeo 4.0 sometime during the next 12 months, which should translate directly into performance improvements for the Smart Video Recorder Pro.

My conclusion? If you're not daunted by the extra $350 and installation hassles, and need video overlay capabilities or 30 frames per second capture, go with the Miro board. In all other instances, the Smart Video Recorder Pro is probably a better choice.

(BetaSP comparison files are PROBTSP.AVI and MIROBT-SP.AVI. VHS comparison files are PROVHS.AVI and MIROVHS.AVI.)

THE NEXT LEVEL

Let's take a look at some advanced capture options to consider when developing your capture strategy.

Targa 2000

Just for fun we thought we would show the results from an EISA-based board that captured with minimum compression. If you've got $5,995 to spare and a fully loaded EISA machine just hanging around, you, too, can achieve this quality level (Fig. 8.12).

All joking aside, from a pure capture perspective, much of the performance offered by this card relates to the bus, not the card itself, since the Targa 2000 uses motion MPEG, the same technology used by Miro and FAST. Equipped with the proper hard drive and controller, an EISA bus can transfer and store as much as 3 MB/second of video to the hard drive. The PCI bus has four times that capacity. This translates directly to video quality.

The Targa 2000 was not without its flaws, producing a wierd diagonal cutting effect on certain frames (see Frame 16:07). Overall, however, the capture quality and color fidelity was quite high.

Figure 8.12 File Produced by TrueVision's Targa 2000

Investing in DRAM

Another way to avoid compression during capture is to capture into huge gobs of memory. At Intel's Indeo compression labs in Oregon, they have several high-end capture stations, one with 64 megabytes of RAM, the other with 128 megabytes. These configurations allow you to capture almost 2 minutes of YUV video without storing to disk.

Typically, you'll want to keep your video clips fairly short, usually under a minute. Out of about 75 clips in the *Video Compression Guide and Toolkit,* for example, none is longer than 60 seconds, and most are less than 30. This means we could capture them all on Intel's runt computer without compression.

How much difference does it make? Actually, on the printed page, you can just spot small additional artifacts above the lip on the left-hand side (Fig. 8.13). However, if you check the real videos at two or three to one magnification, you'll see that

Figure 8.13 Video captured raw in computer with 64 MB of RAM compared to video compressed during capture

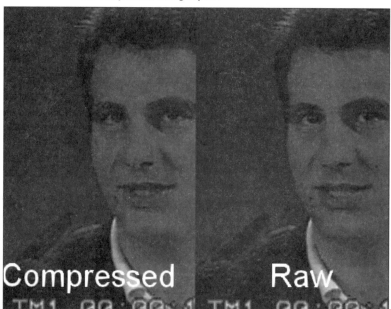

file compressed during capture has many more subtle artifacts (FIG8_13L.AVI and FIG8_13R.AVI).

And finally, the moment you've been waiting for, the absolutely best video capture available on the face of the earth today.

Horizons Technology

Horizons Technology is the company that bought Intel's DVI encoding business back in 1992. As part of the acquisition, they recieved capture stations worth millions.

These stations don't deal with the approximations made by normal capture boards. Instead, they deal with digital analog formats, and convert the digital analog format to digital video format by number crunching, not capture boards.

You may have heard of the D-2 format, or the new Digital BetaCAM decks. These formats store analog information as digital zeros and ones. These zeros and ones are a different breed than those we use on our computers, but they share most other characterisitics. For example, you can edit and copy D-2 formats millions of times without degradation, just as you can copy a PCX file millions of times.

In essence, Horizons Technology does a straight digital-to-digital conversion which is as close to lossless conversion as you can get. You send them your analog video footage on Hi-8, BetaSP or other format, and they dub it to D-2. Then they capture, compress and send you back your video on CD-ROM at a cost of $60.00 per minute of finished video.

We sent Horizons the same BetaSP tape we used for our testing, and they returned raw clips for our preprocessing and compression. Figure 8.14 shows a raw captured clip, just so you can share the awe I felt when I first saw it. We've included a few frames of raw video on the disk in FIG8_14.AVI. Figure 8.15 is the compressed version. It should be against the law to compress video that looks this good, but there it is. Compare

FIG8_15.AVI with other BetaSP versions in Chapters 7 and 8, and you'll get a feel for how good it really is.

Horizons, in San Diego, can be reached at (619) 292–4090.

Figure 8.14 The Holy Grail — HTI's raw video

Figure 8.15 The Holy Grail at 150 kB/second

SUMMARY

1. Table 8-1 shows the four video capture cards that we reviewed, including with a column to use for other boards that you may be considering. Note that they are all ISA-based cards, and that higher performance should be available soon in the form of PCI-based capture cards.

2. The next step upwards is either an EISA based board, the Targa2000 ($5,995) plus the EISA machine, or spending about $4,000 on RAM so you can capture without compression at all.

3. The Holy Grail of Digital Video, the best capture quality we found, was produced by Horizons Technology. At $60.00 a finished minute, this should prove very tempting to developers and corporations seeking the absolute highest possible quality digital video.

Table 8.1 A comparison of video capture cards

Board	Movie Machine Pro	Miro VIDEO DC1 tv	DVA 4000	ISVR PRO	
Company	FAST	MIRO	VideoLogic	Intel	
Performance summary					
15 frames per second	YES	YES	Dropped Frames	YES	
30 frames per second	YES	YES	NO	NO	
Subjective video quality (rank)	4	1	3	2	
Installation ranking	1	4	3	2	
Feature Summary					
Hardware compression	MJPEG	MJPEG	MJPEG	Indeo 3.2	
Bus	ISA	ISA	ISA	ISA	
S-Video Input	NO	YES	Optional	YES	
Hardward Diagnostics	SOME	NO	SOME	YES	
Uninstall Program	NO	NO	NO	YES	
Capture Formats	MJPEG	MJPEG/ RGB	MJPEG/ MSD	YUV/Indeo 3.2	
Limited incoming bandwidth	YES	NO	YES	NO	
Incoming color/brightness control	NO	YES	YES	YES	
Capture Resolutions	Unlimited	Unlimited	Unlimited	limited	
Bundled Software	Premiere	Video Studio	Proprietary	Digital Video Producer	
Other Features	Overlay	Overlay	Overlay	N/A	
Board Configuration	2 boards	one board	two boards	one board	
Price	$940	$849	$2,995	$649	

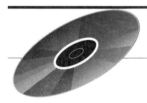

SURVIVING YOUR CAPTURE BOARD INSTALLATION

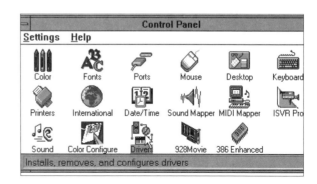

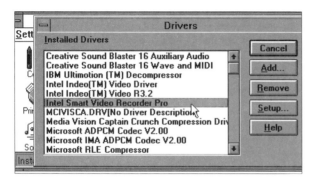

SURVIVING YOUR CAPTURE BOARD INSTALLATION

INTRODUCTION

I've given a lot of thought to the proper attitude to assume when installing a video capture board. Attitude is important, you know, because this task—so simple in the product manual—can be most daunting. My thoughts generally divide into two categories.

The first is the transplant surgeon. Installing a capture board is similar to an organ transplant in that it seems like the computer will do everything in its power to reject the board. This can range from simply refusing to boot after installation, to spurious error messages like "Serious disk error in drive D," to totally incomprehensible messages like "Target code 19008B >16384B changed to 16384B."

The second thought is from the movie *Aliens,* where it seemed like Sigourney Weaver had to kill the space beast seven times before it finally expired. When installing a capture board, you have to be painstakingly, excruciatingly, time-consumingly precise with truly foreign elements such as interrupts, I/O ports and DMA channels or one or more components of the overall capture system—sound card, capture board, graphics board, analog device driver or hard drive—simply won't function.

An engineer buddy of mine swears that it takes two complete days to install and test a new capture board. At first I scoffed—mainly because my first two capture boards, Intel's Smart Video Recorder and VideoLogic's Captivator, practically installed themselves. Then I added a CD-ROM, a SCSI hard drive, the SoundBlaster AWE32, a CD-ROM recorder and

195

MCI drivers for a laserdisk and Hi-8 deck. Now I think two days might be conservative.

It's one thing to install a capture board in an empty computer. It's quite another thing to install one into a multimedia production station.

In June, 1994, just after COMDEX, I collected eight hot new capture boards and allotted a long weekend to test them all. Eight weeks later I finished, but lost three boards along the way. I swear that my computer "rejected" them. After several support calls, consults with hardware and software engineers and many hours of trial and error, we just had to let the boards go. The only thing we didn't try was a faith healer. (Put your hands on the monitor. Heal!!!! Heal!!!!)

During this period I formalized the following concepts about the most efficient way to install a new capture board. While I can't guarantee a trouble-free installation, or that you won't "lose" the patient, I'm pretty certain that you'll shave a couple of hours from your installation.

1. Interrupts, I/O Ports and DMA Channels (Oh, My!)

This section is the poster child for Plug and Play, installation magic supposedly coming with Chicago and PCI. Under Plug and Play, all compliant peripherals will be self-installing—they find their own ports and automatically negotiate any conflicts.

Well, today, most people use a variant of this called Plug and Pray. We put the boards in, boot the computer and hope for the best. You can almost hear Clint say:

"Feeling lucky today, punk? Well, are you? This 80486 VESA Local Bus computer has 16 interrupts. In my excitement over multimedia, I've forgotten how many I've used. Maybe 15. Maybe 16. Feeling lucky today, punk?"

Anyway, in this section we'll discuss ways to avoid Plug and Pray until Plug and Play comes along. If you're an excitement junkie and like the thrill of never knowing whether your computer will boot, you might want to skip this chapter. Otherwise, read on.

WHAT ARE THEY?

Interrupts are signal lines used by peripherals to notify the CPU that it needs attention. You've heard the term "interrupt driven?" For the most part, the host CPU goes merrily along on its way, doing what processors do, until some function in the computer "interrupts" and says, "Hey, I need something." This need could be video capture, a mouse movement, a modem transfer or whatever.

Most capture boards use interrupts, which can be configured either with a hardware switch on the board or totally in software. Either way, your computer only has 16 interrupts, and if you install a new peripheral on an occupied interrupt, you get a conflict, and the new product won't work. The preferred course is to know which interrupts are open and install your new capture card on an open interrupt.

DMA stands for Direct Memory Access line. DMA channels are the data paths to and from system memory. If you install a new device on a busy DMA channel, the old device won't work, and if you're really lucky the computer will lock up. For example, I mistakenly installed a sound card in our development Gateway computer on DMA channel 5, the same as the SCSI board. Instant lock-up. Changed the channel and it was fine. Most computers have eight DMA channels, which is usually more than sufficient.

I/O Ports or addresses are areas of memory used by the CPU to identify the various peripherals or addresses. Your computer has plenty of I/O ports and typically won't run out—problems will only arise if you place two devices at the same address.

So, how to manage all this? After installing our third board, we started charting our interrupts, DMA channels and I/O ports. By keeping track of which were occupied and which vacant, we've saved ourselves hours of debugging time.

HOW DO I FIND THEM?

Several tools are available to help. The lowest common denominator is the Microsoft Systems Diagnostics program, or MSD, which ships with all Windows installations. MSD is an information collection program designed by Microsoft to help localize program bugs and conflicts. It's a great way to gather information about your computer, including the devices sitting on your interrupts.

Before loading Windows, type "MSD" at the command line and the program will load. If you click on the "interrupts" section, the program will identify devices then sitting on an interrupt. Then build a chart like Table 9.1. I'll list what's in our Gateway 2000 development computer, and you list what's in your computer.

Unfortunately, MSD doesn't list DMA channels, although several other programs, including QAPlus from Diagsoft, do. QAPlus came bundled with our Gateway—you might check your hard drive to see if you own a utility that does the same. We track DMA channels in Table 9.2.

I don't know of any utility that tracks I/O Ports—we'll do that ourselves in Table 9.3.

Back to the hunt. Unfortunately, MSD isn't perfect, and it can't identify devices that don't maintain a presence on the

Table 9.1 Charting interrupts

Interrupt Address	Gateway		Your
	MSD	Sleuthing	Computer
Interrupt 0	System Timer		
Interrupt 1	Keyboard		
Interrupt 2	[cascade]		
Interrupt 3	MOUSE		
Interrupt 4	COM3	COM2	
Interrupt 5	LPT1		
Interrupt 6	Floppy Disk		
Interrupt 7	Sound Board		
Interrupt 8	Clock/Calendar		
Interrupt 9	Available	Smart Video Recorder	
Interrupt 10	Available	**Available**	
Interrupt 11	Available	SCSI card	
Interrupt 12	Available	**Available**	
Interrupt 13	80486DX Intel		
Interrupt 14	Hard Disk		
Interrupt 15	Available	**Available**	

Table 9.2 Charting DMA channels

DMA Channel	Gateway Computer		Your
	QA Plus	Sleuthing	Computer
DMA 0		**Available**	
DMA 1	SoundBlaster		
DMA 2	Floppy Disk		
DMA 3		**Available**	
DMA 4	[Cascade]		
DMA 5		SCSI Port	
DMA 6		**Available**	
DMA 7		SoundBlaster	

interrupt line at all times. So you have to sleuth some as well. I've created a separate MSD line in the chart to highlight those we had to track down.

For example, in our Gateway, the SCSI card to the 1.4 gigabyte drive doesn't show up. If you have a SCSI card, it's taking probably an interrupt, a DMA channel and an I/O port. Usually if you peruse your SCSI setup software, it will describe the setup in detail. On our Gateway, the SCSI board takes interrupt 11, DMA channel 5 and I/O Address 330. So I enter these onto the charts.

In addition, I know that my capture board is taking up both an interrupt and an I/O address. However, this won't show up in MSD because the board isn't initialized under DOS. So we take our investigation to the Windows control panel, shown in Fig. 9.1, which is located in your Main Windows Group.

Table 9.3 I/O Adddresses

I/O Addresses	Gateway	Your Computer
2E0-2EF	ATI Graphics Ultra Pro	
310	Smart Video Recorder	
320	SCSI card	
222	SoundBlaster AWE32	
340h	CD-ROM	

Figure 9.1 Control Panel, located in your Main Windows Group

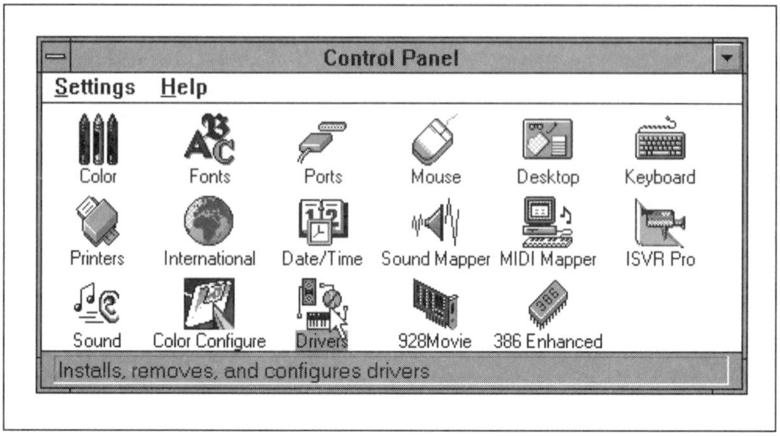

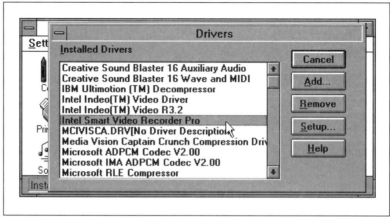

Then double-click on the drivers section to get to the screen shown in Fig. 9.2.

The Drivers dialog box contains all multimedia drivers loaded into your system. It's a convenient place to see what's loaded and what isn't, and depending upon the device, a source of a wealth of information. Let's press the setup for the Smart Video Recorder Pro and see what it tells us (Fig. 9.3).

Figure 9.2 The Drivers Dialog box contains all multimedia drivers loaded into your system

Figure 9.3 Intel Smart Video Recorder Pro Setup Dialog

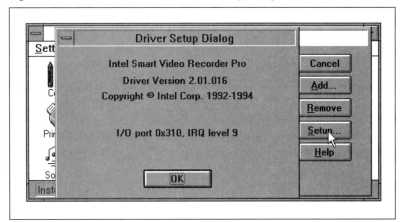

Great sleuthing! Now we know that I/O Port 310 and inter-rupt 9 are occupied, and we mark our charts accordingly. If we decided to replace the Smart Video Recorder Pro, we would probably install the new board right into these parameters.

Now let's hunt for the SoundBlaster AWE32. The Setup Dialog box in the Drivers Section is a dry hole for this unit—not all peripherals list their information there. However, many devices install configuration programs that list their installa-tion locations. Happily, the SoundBlaster Configuration program in the Sound Blaster Windows group is exactly that (Fig. 9.4).

The AWE 32 takes two I/O ports, one interrupt and two DMA channels. Let's plot them in the box.

Graphics boards typically claim first dibs on a set I/O port address. For example, the ATI Graphics Ultra Pro in the Gateway uses 2EO-2EF, whether you like it or not, end of story. So we marked these as taken in our I/O Port chart. Check your graphics card manual to see which ports it takes up.

CD-ROM drives typically take an I/O port and sometimes a DMA channel. Since few software programs identify I/O port addresses, you'll probably have to check your CD-ROM man-ual and installation to find your locations. In the Gateway it's I/O port 340h and no DMA.

Figure 9.4 Sound Blaster Configuration Dialog Box

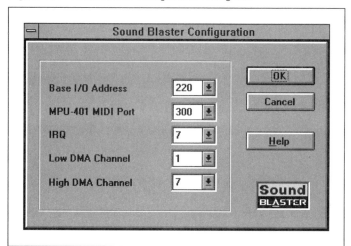

Finally, if you have a modem on your capture station, you'll need to track both the interrupt and comm port. We took out our modem because it conflicted with the comm port that we used to control the laserdisk during step capture. Theoretically, they should have been able to work together, but in practice it didn't work.

These are the primary devices that we have to track. If your computer has other hardware peripherals such as network cards, scanner boards or hardware compression co-processors, you need to find out which resources they consume and add them to the chart.

So where did we end up? Three free interrupts, three free DMA channels. Bring on the peripherals, we'll get them all in!

2. Write It Down

Even after charting the lines in and out of your computer, there's bound to be some trial and error, especially with I/O ports. Get in the habit of writing down every configuration that you try—it will help you rule out bad choices and also help you reinstall the product later if you have to remove it.

3. Configuration Files

In addition to the config.sys file, discussed earlier, several other files contain the essential configuration information about your DOS and Windows system. These are the DOS autoexec.bat file, which is always located in the root of your boot directory (usually C:\) and Windows system.ini and win.ini files, located in your Windows subdirectory (usually C:\windows).

Most capture cards modify some or all of these files during installation. While the more polite installation programs back the files up for you, many don't. Unless you back them up, it's tough to restore your computer to its previous state if the installation fails or should you later decide to remove the device.

Get in the habit of backing up these files before installing any new product. It's usually easiest to copy the file to another name, using the same first name and changing the last name to something that somehow relates to the product you're installing. For example, before plugging in the Smart Video Recorder Pro, we renamed our win.ini and system.ini win.pro and system.pro. Make sure you don't use a last extension of exe, com or bat, just to keep things simple.

4. Two Is a Crowd

It's tempting to get a new sound board, capture card and maybe even a hot new graphics card and really upgrade your computer all at one shot. Unfortunately, even if you do get all the products working, it could take weeks to sort out all the loose bugs and conflicts.

We had just installed our new sound board before we started benchmarking capture boards and hadn't really gotten to know the product. After trying for hours to install the first capture board, we ripped out the AWE32 and returned our faithful 8-bit SoundBlaster to its old slot for the remainder of the testing.

Now that the dust is settled, the AWE32 is back and place and working in perfect harmony with our new capture card. It just took some time for both products to settle in. Had we tried to debug both in real time, it would have taken much longer to return to a steady state.

Finally, never change your video graphics card or drivers. Or at least not casually. It's extremely difficult for these cards to maintain 100% compatibility with all the Windows programs and peripherals, and the competitive nature of the graphics market shrinks beta test time in a rush to release the next great mousetrap. While we like to think that Video for Windows is the center of the universe, it's way behind more mundane applications like word processing and number crunching in the eyes of the graphics cards manufacturers.

If your graphics system is working, resist change. Update your driver for major releases like DCI support, but don't install the latest driver from Compuserve just because it's there. Your capture and compression station is mission-critical, and you can lose days or even weeks trying to restore a configuration hosed by a new graphics card or driver.

5. Always Screw in Your Peripherals

Maybe everybody does this as a matter of course, but I don't, or at least didn't till I started working with capture cards. First thing you do when your capture card won't display video is push that S-Video connector as tight as you can. If the board is loose, you could push it right out, which I've never seen happen but is reportedly worse than what happens if you swim right after eating, or don't brush after every meal.

Same with your sound board. These are both active peripherals, with lots of pushing and pulling. So screw them in during installation.

6. Follow the Manual

If you're playing with video, you obviously know your way around a computer and have probably installed a ton of peripherals and software. If you're like me, manuals are for the other guys, or your mother or whoever. You pride yourself on being able to install a program without the docs. Besides, manuals never help anyway.

Well, this stuff is really hard, and the products are complicated and persnickety. Nothing is worse than spending hours trying to overcome a problem only to find that the answer would have been right in front of your nose had you only looked. If you follow the manual closely during installation, you'll improve your chances of surviving, immensely.

Summary

Well, that's it. Have we scared you off? Is Horizon's $60 per minute compression service looking better all the time?

Nah, you're ready to go! Good luck and good hunting.

From now on we assume you're installed and ready to capture.

PRE-CAPTURE CHECKLIST

1. Defragment Your Hard Drive

When you format your hard disk and start adding programs, all files are stored in continuous sectors on your hard drive. As you delete and add files, DOS starts breaking files up and storing them in noncontiguous disk sectors. For example, if you delete a one-megabyte file and then add a two-megabyte file, DOS might place one half of the new file in the space opened up by the deleted file, and one half in other sectors. Over time, many of the files on your disk are fragmented, and most of the available storage space on your disk is located in noncontiguous sectors.

When you capture video, you achieve the best results by dumping all of the captured data into contiguous sectors. That way, the disk can write continuously, rather than writing, moving to a new sector, writing again, and so on.

Think of a cement truck trying to unload. If it dumps the cement in one large load, like the foundation of a building, it empties very quickly. If it has to unload into forty or fifty wheelbarrows, it will take a lot longer.

When you capture video, if the disk has to seek to new locations to store the newly captured video, you may drop frames. Overall operation will be most efficient if it can just start dumping into continuous sections. For this reason, it's best to start every capture session by defragmenting your hard drive.

Norton Utilities from Symantec contains a defragment utility, which Microsoft licensed for DOS 6.0. Either will do the job. You have to be in DOS to use both programs—you can't be in a DOS window.

To run the Norton version, type "Speedisk" on your command line when in the Norton directory. To run the DOS 6.0 version, type "defrag" from the command line in your DOS directory. Defragmenting your drive can be a lengthy process, depending on the size of your hard drive and how badly fragmented it is. You might try to remember to defragment the night before you plan to capture.

Interestingly, defragmenting also helps video playback. When compressing a file, DOS places video file chunks in available slots all over the disk. During playback, if the file is fragmented, the disk has to work a lot harder to retrieve the file, which can cost you a couple of frames per second. If you're compressing on one station and transferring to another for playback, you should defragment the playback station before or after loading the video files.

Speaking of Norton, it's probably a good idea to install Norton Disk Doctor on your disk and run it each time you boot your computer. The Disk Doctor checks for and repairs file allocation errors and lost chains, makes sure your file allocation tables are correct and keeps a mirror image of disk information in case your drive is accidentally formatted.

With all of the devices and drivers you end up loading on your computer, the inevitable conflicts and the memory-intensive tasks that continuously page information back and forth to disk, inevitably you end up with all kinds of errors on your hard drive. When I installed NDD on my home computer, I found 40 megabytes of lost chains. I locked the barn door and bought Norton after losing my first 1.4 gig drive. I've had some scares, but no problems since.

2. Set File Nomenclature

In the course of capturing, editing and possibly filtering a file for compression, you end up with four or five versions of the same file. Add two to three potential compressions to get it just right, and you've got a bunch of files on disk, all described with eight characters and the surname .AVI. That's just for one final file.

It's always a good idea to develop a naming convention for all generations of video files. We started doing this with the *Video Compression Sampler* since it would have been impossible to manage the thousands of captured, intermediate and final files without one. It's impossible to develop one naming convention that works for everyone, but here's the idea.

First two letters—name/number of the video itself. S1 might be scene one, or V1 might be video one.

Third and fourth letters—describe level or generation of the file.

> *First generation*—use CA (capture) or RA (Raw).
> *Second generation*—after pre-processing use ED (edited) or FI (filtered)
> *Third generation*—when compressing use the name of the codec, like IN (Indeo) or CP (Cinepak)

Fifth and Sixth letters—If you're experimenting to produce the perfect video file, use these letters to describe the parameters. For example, if you're playing with key frames to see which interval gives you the highest quality, you might use these two letters to record key frame interval.

Sometimes the file names are easier to read if you include an underline between the letters. Thus, S1_IN_15.AVI would be the first scene, compressed with Indeo at a key frame setting of 15.

When you have a keeper—or the file you intend to finally use—you might name it something easier to read. For example, you could change S1_IN_15.avi to Scene1.avi.

About file management. I've gotten really conservative about deleting intermediate files, and usually don't start deleting until absolutely necessary. Pretty much every time I delete a filtered or raw file, it seems like I need it again. We obviously do a bit more experimenting than most readers, but you may find the same thing.

3. Plan Your Directory Structure

Even the best file-naming convention doesn't help when you've got 5,000 files in one subdirectory. Think about logical ways to segment the capture information, and then use separate subdirectories where you can.

4. Keep a Log

I'm the world's worst record keeper, and resisted keeping a log for about eight months. But between the time you capture a file and compress the final version, you'll deal with the following parameters:

(a) Color/brightness settings on the capture card.
(b) Time codes or frame numbers on the analog source.
(c) Color/brightness settings in Adobe Premiere or other editor.
(d) Filter settings.
(e) Compression settings.

It's bad enough when you have to start over because the boss thinks the videos makes her look like a Martian. It's really demoralizing when you have to start from scratch without a record of what worked before and what didn't.

5. Get Comfortable with Your Video Input Controls

In terms of pure video quality, the most valuable time you'll spend with your capture board is with analog video input controls. As you recall, these adjust the color and brightness of the incoming video. While some of the capture boards come close, default settings don't work 100% of the time, and adjustments are necessary to maximize video quality.

In the heat of the capture, you focus on getting the video to disk, and your internal quality controls become more lax. Later, during preprocessing and compression, you get more critical and start thinking gosh, this would look so much better if only it was brighter, or darker, or less green or more green. Then you might talk yourself into starting over, or trying to adjust the color in Premiere.

The key to one-step video quality lies with the analog input controls. Get to know them early.

6. Don't Delete More Than You Can Chew

This is kind of a silly one, but whenever I get ready to capture, I start deleting files left and right to clear the space. It's all backed up, mind you, but it's a pain to restore and I never end up capturing anywhere near as much as I thought I would.

The weekend before my capture board *grand prix* I cleared 1.2 gigabytes of space that took hours to restore. My ultimate harvest for the day? About 100 megabytes.

SUMMARY

1. Charting your I/O converts Plug and Pray into a much less frustrating routine, whether for capture boards or other computer peripherals. Interrupts, I/O ports, and DMA channels sound scary, but it's not rocket science and you *can* figure it out.

2. Always install new peripherals one at a time. This will simplify the overall process greatly.

3. Defragmenting your hard drive is the key to real-time capture performance. Defragment before every major capture session.

4. Create a standard file nomenclature before you start capturing. Make up unique identifiers for the various clips you'll be capturing, as well as all stages from capture to compression. When you have 6,000 similarly named AVI files on your disk, it's too late.

5. Keep a log of all your work. Track capture settings, start and stop points, and filter and compression settings. This will help you recreate video files, which inevitably you will have to do.

6. Get comfortable with your capture board's input color controls. Proper use of these controls guarantees maximum image quality. Improper use guarantees you'll have to redo the videos.

VIDEO CAPTURE

10

VIDEO CAPTURE

In this chapter we start by counting down to capture, reviewing the various VidCap controls that implement our capture decisions and discussing critical capture parameters such as frame rate and video resolution. All supported by sample videos and abundant statistics, of course. Then we move to the capture itself, reviewing topics like how to configure computer controlled analog sources, working with the MCI controls and how to step capture.

   This section is very hands on, best used when actually doing the work. So clear some space for me on your desk, right there near the keyboard, and let's get down to work. It's time to capture some video.

CAPTURE CHECKLIST

1. Set Capture File

When you load VidCap for the first time, it tells you to set the capture file. Once set, VidCap stores the video to that same file every time you capture unless you change the file name. If you capture 10 files, and don't change the name, you'll end up with only one file—the last—and the fruits of your previous captures will be random unrecoverable zeros and ones on your drive.

Figure 10.1 VidCap's Smart Icon bar. From left to right, it's Set Capture File, Load VidEdit, Preview Video, Overlay Video (gray), Capture Frame, Capture Frames, Capture Video and Capture Palette

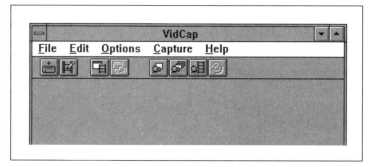

Remember the "IT'S THE ECONOMY, STUPID" sign in Bill Clinton's campaign headquarters during the 1992 election? Whenever we're about to capture a bunch of files, we always put a yellow sticky note up on the computer monitor that says:

CHANGE THE CAPTURE FILE, STUPID

You might consider doing the same—even if you're a Republican or Perotista. It won't stop you from writing over files by mistake, because we all do it. It may cut down the number, though.

To set the capture file, you can either use the File/Set Capture File command, or push the "Set File" smart icon where shown in Fig. 10.1.

2. Set File Size

When you set a new capture file, the software prompts you to set file size (see Fig. 10.2). What this actually accomplishes was the subject of a hotly contested debate at a seminar I taught last July.

When you set the value at 60 megabytes, you reserve 60 megabytes of disk space for the capture file. If the actual video

Figure 10.2 VidCap's Set File Size box lets you allocate a capture
working area

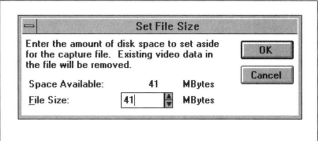

captured is smaller, say 10 megabytes, the capture file still
takes up the original 60. It doesn't take more than a few 60-
megabyte chunks to fill up even a large disk drive, so I quickly
started setting the capture file to one megabyte without problems. This procedure was also recommended in a white paper
distributed by Intel.

At the seminar, one participant said that by reserving the
space, you reserve contiguous sectors of unfragmented storage.
He recommended capturing into the 60 megabyte storage
space and then using the Save As command to save the file to
another name. This stores the file at its actual size, and avoids
cluttering your drive with allocated, but unused storage space
in these capture files.

I didn't think that reserving the space would guarantee
access to unfragmented sectors. In other words, I saw nothing
in the help files or other documentation that guaranteed
that all of the space reserved would be contiguous and
unfragmented.

Here's what Microsoft's help file says:

> Before capturing video sequences, you must identify a cap-
> ture file. The capture file sets aside an area of your hard
> disk to receive captured video and audio data. By ensuring
> that the capture file exists in a contiguous area on your
> hard disk, you can allow the highest possible disk-write
> speeds and reduce the chances of losing data due to disk
> seeks.

When creating a capture file, specify a size large enough to hold the entire video sequence. If you run out of room while capturing, VidCap will attempt to enlarge the file to hold the captured data. However, enlarging a capture file might cause the file to become fragmented and might also cause one or more frames to be dropped.

Hint: You can use a capture file multiple times; when you finish capturing video, just save the video sequence to a different filename. You can also set aside multiple capture files.

Note that Microsoft doesn't say that the capture file sets aside a "contiguous" area on your hard drive, it says that "you" have to ensure that the area is contiguous.

When I got back to the office, I performed the following experiment.

(a) Set a capture file on the lab computer to observe how much space was present. Total space available was about 46 megabytes.

(b) Deleted a bunch of small video files, and increased the available space on the drive to over 180 megabytes.

(c) Set a new capture file and reserved all 180 megabytes of storage.

(d) Loaded Speedisk, which reported that 95 percent of the drive was unfragmented, but showed that the available storage was spread across the entire disk.

If VidCap reserved only contiguous, unfragmented sectors, I couldn't have reserved anywhere near the 180 megabytes allowed by the program—there simply wasn't a 180 megabyte contiguous unfragmented block available on the drive. So while reserving space may grab contiguous sectors when available, it doesn't ensure that you'll get them.

Probably the safest course is to defragment your hard drive and then set a permanent capture file to use during the capture session. Each time you capture, use the Save As command to save the file to another directory. If you start with an unfragmented disk, this would appear to ensure that the capture area itself remains unfragmented.

If you chose this route, change the sticky note recommended in the previous section from "Change the Capture File,

Stupid" to "Save the File, Stupid!" If you capture two files in sequence, the first file will be written over and unrecoverable.

Note that all of this becomes much less important if you step-frame capture rather than capturing in real time. Step-frame waits until the computer is ready to send the next frame, which includes storage time, so storing in noncontiguous regions doesn't affect capture performance. However, remember that fragmented compressed video files don't play back as fast as unfragmented files. Overall, it's a pretty good idea to defragment pretty regularly no matter which way you capture.

3. Audio Format

CHOOSING THE RIGHT AUDIO PARAMETERS
Audio format determines the quality of your audio, and audio parameters are a design decision that should be set early in the development process. Obviously, audio adds to the bandwidth of your captured video, meaning that higher-quality settings add load that the may cause you to drop frames during capture. For this reason, you should capture at the audio setting you intend to use, never higher. Table 10.1 illustrates the data rates associated with sampling frequency, channels and sampling size.

Here's what these parameters mean. You're probably familiar with the difference between mono and stereo—stereo carries two channels and is twice as large. Sample size relates to the amount of storage used per unit of analog audio. As you would expect, 16-bit requires roughly twice the data rate of 8-bit.

Table 10.1 Audio data rates at specified audio parameters

	Channels	Mono Audio		Stereo Audio	
Frequency	**Sample Size**	**8-bit**	**16-bit**	**8-bit**	**16-bit**
11.025 kHz		11KB/S	22KB/S	22 KB/S	44 KB/S
22.05 kHz		22 KB/S	44 KB/S	44 KB/S	88KB/S
44.1 kHz		44 KB/S	88 KB/S	88 KB/S	176KB/S

Finally, frequency relates to the rate of sampling performed on the audio. The higher the sampling rate, the higher the fidelity of the captured audio. By way of reference, 44.1 kHz is used for audio CDs, 22.05 kHz sounds about the same as AM radio, while 11.025 kHz is about the same quality as a telephone line.

As you can see, at the higher-quality sound settings, the results are somewhat alarming, especially if you're working with CD-ROM products. Choosing the right combination relates primarily to the sound system on the target computer. Obviously, if you can't hear the difference, the additional quality isn't worth the bandwidth.

Whatever combination you chose, make sure you test on a computer with a very low-end sound board, which often yields some interesting results. For example, I've experienced a fuzzy sound called "white noise" when playing 16-bit audio on an 8-bit sound systems. Other developers have reported audio distortion when high-quality audio is played on lower-quality sound systems.

For these reasons, and to save bandwidth, most CD-ROM publishers seem to use either 22 or 11 kHz, 8-bit mono audio. All of the videos on the CD-ROM use 8-bit, 11 kHz mono audio.

AUDIO FORMAT CONTROLS

Set audio format with the dialog box shown in Fig. 10.3. Always, always, always check audio levels as shown in Fig. 10.4 to see if you're receiving audio during capture. If the line doesn't jump to the right, audio is not getting through and won't be there during capture.

VidCap doesn't control audio input as it does video input. Instead, audio control is usually handled by the software that comes with your sound boards. My first Sound Blaster had a really simple control—a nice knob on the card that was tough to reach but simple to understand. Turn it to the right to make it louder, to the left to make it softer.

My new SoundBlaster AWE32 comes with software designed after the controls found in sophisticated analog editing suites. This is great if you work for Arista Records but

Figure 10.3 Dialog box for setting Audio Format

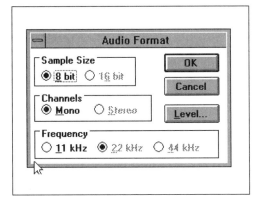

Figure 10.4 Audio level showing incoming audio, indicating that it's safe to capture

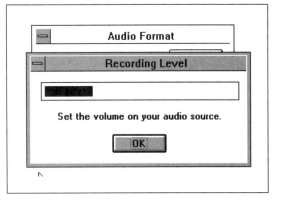

somewhat confusing if the most sophisticated analog device you've controlled is your car stereo. Ah, my curmudgeon is showing—the software worked fine after a couple hours of experimenting, and the sound was truly AWEsome.

Basically, this is my way of saying that you're on your own—each software package works differently, and there are too many to detail here and not enough time to learn them all. Whatever program you own, however, check the dialog box shown in Fig. 10.4 frequently or you could inadvertently end up in silent movies.

4. Video Resolution

PRIMARY TARGET SYSTEM

Video resolution is one of those **GLOBAL** issues to be considered during the design stage rather than late in the game. It also must be analyzed along with frame rate, since capture resolution affects display rate. Both issues evolve around the concept of *Primary Target System,* loosely defined as the system towards which your product is primarily targeted.

As we saw in Chapter 5, video performs differently according to processor and bus type, and even distribution medium.

Chapter 6 showed us that different codecs are appropriate for different video types, and also perform differently.

Now we have to process that information into application-specific concrete compression parameters. The starting point is the hardware configuration of your primary target system.

In some instances, isolating your primary target system is easy. Maybe you have a dedicated sales force equipped with the same class of laptop, or three or four kiosks with the same hardware. On the other hand, if your application is campus- or corporate-wide or you're shipping into commercial markets, the decision is much more difficult.

My crystal ball isn't clearer than anyone else's. Here are some observations you may want to consider.

USER PREFERENCES

I recently worked with about 40 educators at a seminar. We showed various videos in different resolutions, and asked, "Which looks better, the one on the right or the one on the left?" Anyone listening in must have thought that I was an eye doctor. Here's the general result.

Talking head videos (audio/video synchronization required)— For talking head videos, most preferred synchronization over window size. That is, most liked a 240x180 video that played at 15 frames per second (fps) better than the 320x240 video that played at 12–13 (fps). In low-motion sequences, still-frame quality was also important, which once again pointed towards the smaller resolution (compare talk240.avi with talk320.avi in chap_10 subdirectory).

Moderate motion (no audio/video synchronization)—Most preferred window size over frame rate. That is, most liked 320x240 at 10 fps over 240x180 at 15 fps. Moderate motion didn't appear jerky at this frame rate, and the bigger picture illustrated the subject of the video more clearly (Compare mod240.avi with mod320.avi in chap_10 subdirectory).

High motion (no audio/video synchronization)—For higher-motion videos, smoothness seemed more important that window size. Most preferred a 240x180 video that played at 15 frames per second to a 320x240 video playing at 10–12.

(Compare himo240.avi with himo320.avi in chap_10 subdirectory.)

In general, the educators seemed to focus more on their newer computers than the installed base of older systems. They seemed willing to let the users of older systems suffer poor performance so that those with newer systems could really leverage the performance. Thus, their primary target systems were primarily 80486 local bus systems rather than 80386 and ISA 486s.

General corporate—It feels like most corporate training and MIS coordinators are concerned with their installed base, and want their videos tailored to a broader group. Thus, their primary target system is focused around the ISA-based 80486.

Consumer titles—Most consumer titles still assume a single-spin CD-ROM drive (150 kB/s transfer rate), but we're starting to see more and more 320x240 videos, which assume a fairly high performance computer.

QUICK ANSWER

Low motion:	240x180
High motion:	320x240

Rule: Select a frame rate/compression resolution combination that allows all frames to be displayed during playback on your primary target system. This maximizes video quality and display rate.

Important: Always capture at your target resolution. Video capture cards use sophisticated interpolation and filtering to minimize scaling artifacts. VidEdit scales through simple sub-sampling which produces noticeable pixelation and other artifacts (Fig. 10.5).

OPERATION

Tables 10.2 and 10.3 show how Cinepak and Indeo perform on a sampling of target computers. The results shown are for Indeo 3.1, not Indeo 3.2. Preliminary tests we performed with the Beta version of Indeo 3.2 show it to be about 2–3 frames

Figure 10.5 Image captured at 240x180 (left) compared to image captured at 320x240 and scaled by VidEdit to 240x180. Compare smooth collar and clear numbers on left to ruffled collar, distorted numbers and slightly distressed look on right.

per second faster than the previous version, but still slower overall than Cinepak.

We were impressed with both codecs' performance on lower-end computers—so long as you stay in 8-bit mode.

Table 10.2 Display rate matrix, Cinepak—action sequence

		\–160x120–\		\–240x180–\		\–320x240–\	
Primary Target System		**10 fps**	**15 fps**	**10 fps**	**15 fps**	**10 fps**	**15 fps**
386/33 ISA	8-bit Hard Drive	10 fps	15 fps	10 fps	15 fps	10 fps	15 fps
	24-bit Hard Drive	10 fps	15 fps	2 fps	1 fps	1 fps	1 fps
486/33 ISA	8-Bit CD-ROM	10 fps	15 fps	10 fps	15 fps	10 fps	14 fps
	Hard Drive	10 fps	15 fps	10 fps	15 fps	10 fps	15 fps
	24-bit CD-ROM	10 fps	15 fps	3 fps	2 fps	1 fps	1 fps
	Hard Drive	10 fps	15 fps	4 fps	3 fps	1 fps	1 fps
486/33-VLB	8-bit CD-ROM	10 fps	15 fps	10 fps	15 fps	10 fps	15 fps
	Hard Drive	10 fps	15 fps	10 fps	15 fps	10 fps	15 fps
	24-bit CD-ROM	10 fps	15 fps	10 fps	13 fps	10 fps	13 fps
	Hard Drive	10 fps	15 fps	10 fps	14 fps	10 fps	14 fps
486/66-VLB	CD-ROM	10 fps	15 fps	10 fps	15 fps	10 fps	15 fps
24-bit	Hard Drive	10 fps	15 fps	10 fps	15 fps	10 fps	15 fps
Pentium-PCI	CD-ROM	10 fps	15 fps	10 fps	15 fps	10 fps	15 fps
24-bit	Hard Drive	10 fps	15 fps	10 fps	15 fps	10 fps	15 fps

Table 10.3 Display rate matrix, Indeo 3.1 talking head sequence

		\–160**x**120–\		\–240**x**180–\		\–320**x**240–\	
Primary Target System		10 fps	15 fps	10 fps	15 fps	10 fps	15 fps
386/33 ISA	8-bit Hard Drive	10 fps	15 fps	10 fps	3 fps	2 fps	1 fps
	24-bit Hard Drive	1 fps	1 fps	0 fps	0 fps	0 fps	0 fps
486 ISA	8-bit CD-ROM	10 fps	15 fps	10 fps	15 fps	10 fps	11 fps
	Hard Drive	10 fps	15 fps	10 fps	15 fps	10 fps	13 fps
	24-bit CD-ROM	10 fps	15 fps	3 fps	2 fps	1 fps	1 fps
	Hard Drive	10 fps	15 fps	4 fps	3 fps	1 fps	1 fps
486/33-VLB	8-bit CD-ROM	10 fps	15 fps	10 fps	15 fps	10 fps	14 fps
	Hard Drive	10 fps	15 fps	10 fps	15 fps	10 fps	15 fps
	24-bit CD-ROM	10 fps	15 fps	10 fps	12 fps	6 fps	7 fps
	Hard Drive	10 fps	15 fps	10 fps	14 fps	8 fps	9 fps
486/66-VLB	CD-ROM	10 fps	15 fps	10 fps	15 fps	10 fps	14 fps
24-bit	Hard Drive	10 fps	15 fps	10 fps	15 fps	10 fps	15 fps
Pentium-PCI	CD-ROM	10 fps	15 fps	10 fps	15 fps	10 fps	15 fps
	Hard Drive	10 fps	15 fps	10 fps	15 fps	10 fps	15 fps

However, on ISA machines, performance quickly ground to an ugly halt at higher color depths. For example, an ISA 486 test machine that displayed 320x240 videos from both codecs at or close to 15 fps in 8-bit mode only managed about 1–2 frames per second in 24-bit mode. This relates to the fact that the second bus transfer, from main memory to the video card, is three times larger in 24-bit than in 8-bit mode.

It may be worth recommending that ISA bus owners play all videos in 8-bit mode. This places a premium on effective palette management, which is kind of a pain, but well worth the trouble if selling into these markets.

Playback Medium—As we learned in Chapter 5, retrieving data from a CD-ROM is more demanding than retrieval from a hard disk. As the two charts show, this can cost 2–3 frames per second.

Video resolution is selected in the VidCap control shown in Fig. 10.8 (p. 226).

5. Video Format

This control is specific to the individual capture boards. The Smart Video Recorder, shown in Fig. 10.6, offers two capture formats, raw YUV-9 and compressed Indeo 3.2. Most other capture boards also offer a compressed and raw format, although often the raw format is 8, 16, or 24-bit RGB format, or all three, rather than YUV-9. Since converting the video to YUV-9 format is the first step most codecs take during compression, RGB offers no inherent advantages and RGB files are two to three times larger than YUV.

FORMAT—STEP-FRAME CAPTURE

When step-frame capturing, always capture in the raw format. If the capture boards offers both YUV and RGB formats, capture with both and see which offers the best color fidelity. If there's no difference, use the YUV and save the file space.

There really shouldn't be any noticeable difference between YUV and RGB formats, but YUV was clearly off-color on several boards that we've tested. I should point out that the Smart Video Recorder was *not* on the these boards and offered excellent color fidelity in YUV-9 format.

FORMAT—REAL-TIME CAPTURE

In most instances, capture in compressed format when capturing in real time. The only exception is where your capture sta-

Figure 10.6 Capture formats offered by the Smart Video Recorder

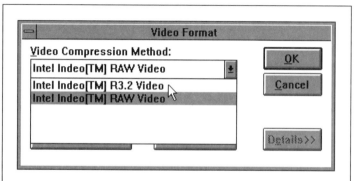

tion is either so fast or has so much RAM that you can capture and store all the target frames to disk without compression.

When capturing in compressed format, always capture at the highest-quality setting offered by the board. Don't try to capture down to your target data rate in one pass. While this usually forces a double compression, overall your video will look much better. Here's why.

Most capture boards, including the Smart Video Recorder, apply only intraframe compression during capture. All captured frames are key frames, and no interframe compression occurs.

While intraframe compression works well at high data rates, we know that overall, interframe compression is more efficient. Capturing at the highest possible quality and then recompressing down to your target data rate lets you bring interframe compression to bear, and improves quality.

Figure 10.7 compares two video files, one captured at about 170 kB/second, the lower limit of the Smart Video Recorder Pro, the other at about 370 kB/second, which was the upper

Figure 10.7 Video files captured at high and low quality settings

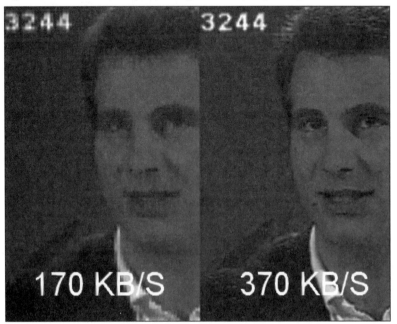

quality limit for 15 frames per second on our capture station. Both were preprocessed and compressed identically. As you can see, the file captured at 370 kB/second is much clearer than the other.

Unless you're capturing for hard-drive playback, where you can accept data rates in the 300 plus range, your best strategy is to capture at the highest possible quality your computer will allow and then post-process and compress to the final data rate.

Both video files are contained in the chap_10 subdirectory under Fig8_7l.avi and Fig8_7r.avi.

To access the quality control, press the "details" button shown in Figure 10.8. The Smart Video Recorder lets you meter the compressed input by controlling either quality or data rate (Fig. 10.8). Try both options and see which produces the highest-bandwidth file.

Figure 10.8 Controlling compression during capture by selecting data rate or quality

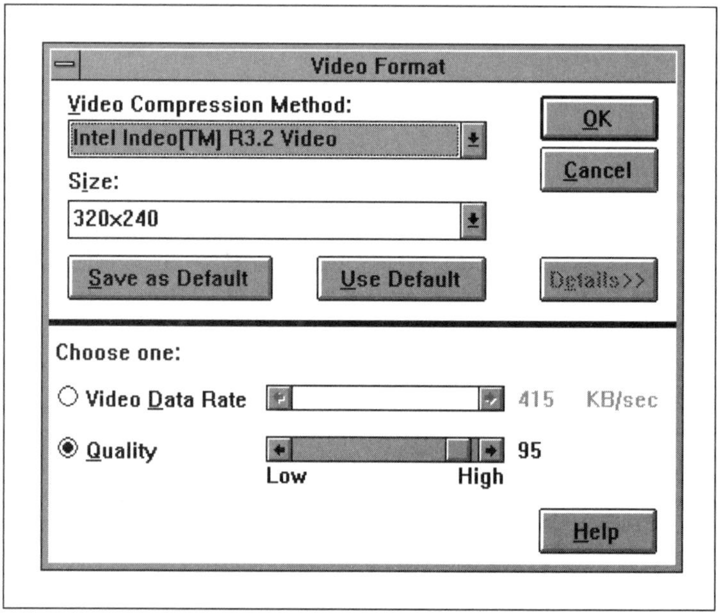

Video Format
Video Compression Method:
Intel Indeo[TM] R3.2 Video
Size:
320×240
Save as Default   Use Default   Details>>
OK   **Cancel**
Choose one:
○ Video Data Rate   415   KB/sec
● Quality   95
Low   High
Help

DROPPED FRAMES DURING CAPTURE

The only time to consider capturing at less than maximum quality is if you're dropping frames during capture. Dropping quality should always be your last resort, however, after trying all of the following options:

(a) Defragment your hard drive (again).
(b) Change to 8-bit (256-color) video mode. This reduces the computer's overall workload allowing more CPU cycles for capture, transfer and storage.
(c) Don't use the "preview" option during capture, since it takes CPU cycles to update your video card.
(d) Unload all extraneous memory resident programs.

6. Video Source

SELECTING YOUR ANALOG SOURCE

Like most cards, the Smart Video Recorder offers both composite and S-Video input jacks, and at times both may be hooked to live video sources. A dialog box lets you select which input signal to capture (Fig. 10.9).

Notwithstanding this control, however, it's probably not a good idea to have two simultaneous live feeds entering the video card. We've noticed interference on several occasions in this situation. If you have two decks hooked into your video card, it's probably good practice to shut one down when capturing with the other.

Several capture programs, including VideoLogic's DVA4000 program, enable a custom selection when capturing from a VTR (video tape recorder), or VCR (video cassette recorder) as opposed to directly from a camera or other source (Fig. 8.4). VTRs can produce an irregular synchronization signal which manifests in rolling, hooking or jittery displays. In theory, when you check the VCR or VTR box, you enable compensation called time-based correction, which cleans up the signal.

Figure 10.9 VidCap's Video Source dialog box where input source is
selected and the incoming video is edited

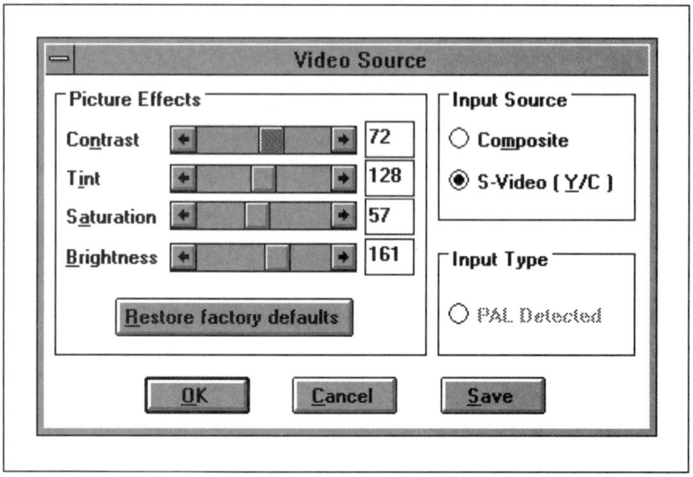

In operation, check the box on and off and see how
it affects your incoming video signal from whatever source. If
it cleans it up, great. If not, click it off and don't worry about
it.

ADJUSTING COLOR AND BRIGHTNESS

This dialog box also contains color and brightness adjustments
when offered by the capture boards. Most of the time, the
default settings simply won't work—you'll probably have to
customize the settings for each class of videos that you capture.
Accept it, get used to it, don't fight it, or you'll end up redoing
a bunch of video files.

Unfortunately, selecting the optimal settings is a real pain.
It's subjective, it's time-consuming and the controls are
obscure. I mean, can anybody meaningfully explain what hue
and saturation are?

If your analog source is frame-accurate, like a laserdisc, it's
somewhat easier to adjust because you can freeze the frame on
screen and adjust the controls until you get it right. With other

decks, you've just got to roll the video and experiment in real time.

Don't expect the captured video to look like the video you modified on screen—the color always seems slightly different. This is especially so when capturing with an overlay card. Analyze small chunks of the actual captured video to make sure it's acceptable.

Don't expect your settings to work with all video segments on the tape, even if you don't change scenes. Sometimes color and/or brightness changes over the life of the tape, or maybe it was the lighting adjustment that you did halfway through the shoot. Either way, you may have to change the settings to keep your video looking consistent.

If you've got disparate footage on the same tape, like outdoor and indoor shots, you should also check your parameters when adjusting to the new footage. The "indoor" settings just may not work for the great outdoors.

You should also calibrate when changing tapes. For example, if your video shoot took three BetaSP tapes, check your settings between each one. You can calibrate the new video against the old by loading a file in VidEdit and keeping it on-screen as a guide.

Once again, all this work can be a real pain. However, when the colors aren't just right, your video won't look good. While you can modify colors and brightness in Premiere, it's better to get it right the first time.

PREVIEW VIDEO/OVERLAY VIDEO
Use these options, both available by menu and toolbar, to view the incoming video. Preview video is available in virtually all instances. Overlay is available only when the feature is supported by both the capture card and video card, and when the necessary hardware and software is installed.

When working with preview enabled, mouse responsiveness will slow down to a crawl as the CPU struggles to keep up with 30 frames per second of incoming video. When this occurs, it's usually easier and faster to work with the keyboard rather than the mouse.

TROUBLESHOOTING THE INCOMING SIGNAL

At this stage you should have a clean video signal displayed on your monitor in either overlay or preview mode. Here are some problems that may occur, and some possible solutions.

(1) **No Video, Period!**—This can be caused by everything from a capture card that's not properly installed to a cable that's too loose. This is often the most frustrating experience of all, since you really don't know where to start. Having an NTSC monitor really helps, because then you can tell if your analog source is working. Here's some things to try:

(a) *Verify software setup*
- Make sure you've selected Preview Video.
- Make sure that you've selected the right input source (S-Video/Composite) and broadcast standard (PAL/SECAM/NTSC).
- Toggle the VTR/VCR button to see if that makes a difference.
- Make certain that the driver is properly loaded. Double-click on Main Control Panel—Drivers and see if the capture board is listed. If it isn't, rerun the installation program.

(b) *Verify hardware setup*
- Check cable connections to the deck. Make sure output is selected.
- If you have an NTSC monitor that's receiving the video signal, switch connectors and see if that brings the board to life. If it does, you'll know that your board is properly installed and that you may have a faulty S-Video cable.
- If you don't have an NTSC monitor, try the composite signal (assuming you've been using S-Video) and see what happens.
- Shut the computer down and make sure that the analog connectors are properly seated in the capture board. With my first capture card, properly seating the video cable

required a lot of pressure, which got me in the habit of
screwing in the board.

- Change cables.
- Reverify your hardware installation by checking the manual. Some boards get pretty tricky.
- Check the user manual for additional advice.

(2) *Video Jittery or Otherwise Distorted*—This almost
always seems to be attributable to an incorrect software setup.
Check to make sure that you've properly selected the right
input source and toggle the VCR/VTR button.

(3) *Video Mottled or Gray-scale*—This almost always seems
to relate to a software setting as well. Some YUV formats can
look pretty funky on screen with greenish Martian types of
effects the most frequent. Some boards capture 8-bit video in
gray-scale.

Also check and make sure you've selected the proper input
source. Sometimes when Composite input is checked but the
input source is S-Video, the video appears as gray-scale.

7. Frame Rate

QUICK ANSWER

15 frames per second (fps)—consider 30 fps when your target system includes a generic video co-processor such as
VideoLogic's Movie928. Consider 10 fps for moderate-motion
clips if you prefer a larger video window over the appearance
of synchronization.

Exceptions—When creating video files from ScreenCAP, the
Video for Windows utility, or converting animations or
morphing, lower frame rates are often preferable. We used
between 5 and 10 fps for screen captures in the *Guide to Video
Compression*.

Dependencies: compression resolution—Do not capture
or modify frame rate without first selecting compression
resolution.

DISCUSSION

Select capture frame rate within the control shown in Fig. 10.12. The frame rate is the number of frames per second in the compressed video stream. This differs from the display rate, which measures how many frames per second are actually displayed during video playback.

Most video is composed of 30 discrete frames per second. During capture, you digitize these discrete analog frames—you don't sample from a continuous analog stream. For example, when capturing at 24 fps, you drop every sixth frame—you don't capture 24 evenly spaced frames from the one second of video.

When capturing at frame rates not evenly divisible into 30, you introduce irregular intervals into the video stream. At 20 fps, for example, every third frame is dropped, which appears less smooth than 15 fps even though the actual number of frames per second displayed may be higher. For this reason, 10 and 15 fps, which divide evenly into 30 fps, are the preferred capture frame rates.

Audio is typically captured along with the video. Usually, the audio is segmented into the number of video frames captured and then interleaved with each frame. For example, if you capture at 15 frames per second, the audio is divided into 15 equal chunks, each interleaved with their respective video frames.

Always capture at your targeted frame rate. While you can modify the frame rate after capture, this usually distorts the frame sequence. For example, if you capture at 15 fps and later drop to 10 frames per second, VidEdit drops every third frame, creating a video sequence that's different than had you originally captured at 10 fps. This is illustrated in Table 10.4.

FRAME RATE OPTIONS

10 frames per second—Use 10 fps if you prefer larger-sized video windows. Because the same video bandwidth is spread over fewer frames, each individual frame is usually of higher quality. However, at 10 fps, most talking head videos don't

Table 10.4 10 fps video originally captured at 10 fps, and captured at 15 fps and adjusted to 10 fps

	\— —		Com	pres	sed	fram	es	— —	—\	
	Frame 1	Frame 2	Frame 3	Frame 4	Frame 5	Frame 6	Frame 7	Frame 8	Frame 9	Frame 10
Actual Video frames										
Captured at 10 fps	1	3	6	9	12	15	18	21	24	27
Captured at 15 fps and reduced to 10 fps	1	3	7	9	13	15	19	21	25	27

look synchronized and most action sequences appear choppy. 10 fps is most often used with animated sequences, morphs and screen captures, where synchronization is not as critical.

15 frames per second—Offers the best blend of quality and display rate and is the most commonly used frame rate for real-world videos.

30 frames per second—The Holy Grail of digital video. Unfortunately, no current software codec produces high-quality video at 30 fps at 320x240 resolution at 150 Kb/second. Compare the single-frame quality of a 15 frame per second video with its 30 frame per second counterpart in Fig. 10.10 and you'll get the idea. For this reason, while Pentium systems

Figure 10.10 Video compressed at 30 fps (left and 15 fps, both at a bandwidth of 150 KB/second

Table 10.5 Display rate matrix for files compressed at 15 and 30 fps; all files compressed to 150 kB/s

	\——Frame	Rate——\
	15 fps	30 fps
Display Rate—Indeo Talking head (486/66 VLB 24-bit)	15 fps	23 fps
Display Rate—Cinepak Action (486/66 VLB 24-bit)	15 fps	30 fps
Display Rate—Cinepak Action (486/33—ISA 8-bit)	15 fps	2 fps (that's right, 2)

can play back 320x240 at 30 fps in many instances, video quality is poor.

In addition, most 486-based computers can't decompress and display 30 fps. The overhead associated with the additional frames often decreases the display rate to below that produced by 15 fps files, detracting from perceived video quality and audio/video synchronization (Table 10.5). Moreover, the frames actually displayed look worse than the 15 fps frames, since twice the compression must applied to the 30 fps file.

VIDEO CO-PROCESSORS

While compression resolutions of 160x120 produce higher-quality results at 30 fps, they typically can't be decompressed and scaled to 320x240 without a video co-processor such as the Movie928. If you're targeting systems with these devices, you may get the best results by compressing at 30 fps at 160x120 and scaling to 320x240 during playback. The co-processor typically ensures that you'll display all 30 frames per second, and video quality is superior to 320x240 files compressed at 30 fps, particularly in high motion sequences (see Fig. 10.11).

Remember that the Movie 928 doesn't accelerate video played back at its normal resolution and produces a display rate similar to the ATI Ultra Pro for 320x240 videos played back at 320x240. This is because the current implementation only scales *after the video is on the card*; it doesn't assume functions otherwise performed by the CPU. See Table 10.6.

Figure 10.11 Interpolated 30 fps video with the Movie928 gives 320x240 resolution at 160x120 quality

8. Miscellaneous Options

Now we're getting close to the actual capture. The Capture Video Sequence dialog box shown in Fig. 10.12, is where you implement your Frame Rate Selection and take final passes at audio and video capture parameters. But first, the Enable Capture Time Limit option.

ENABLE CAPTURE TIME LIMIT

By selecting this option in the dialog box, you can set a time limit for the capture and have the software stop the capture

Table 10.6 Display rate matrix illustrating performance boost from video coprocessor. Tests performed on a 486/66 VLB bus computer with 8 MB of RAM. Files were compressed to 150 kB/s and played back from a Texel 2x CD-ROM drive.

	Display Rate	
	ATI Ultra Graphics Pro	Movie 928
Display rate—160x120 @ 30 fps Indeo file zoomed to 640x480	2 fps	30 fps
Display rate 320x240 @ 30 fps Indeo file at 320x240	22 fps	22 fps

Figure 10.12 Capture Video Sequence dialog box

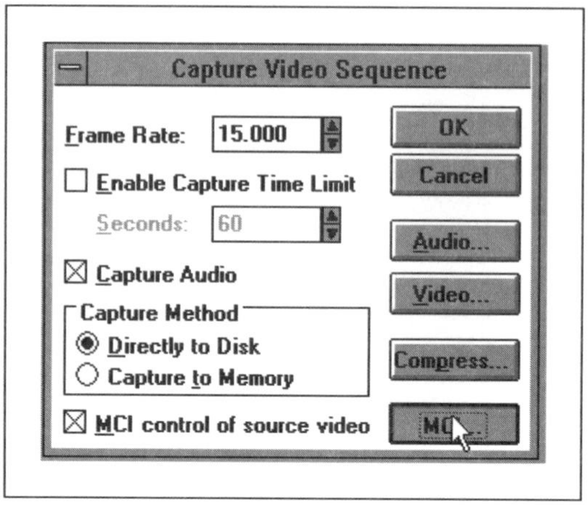

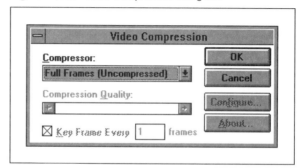

automatically. It's difficult to imagine when you would want to use this option, but there it is.

The other buttons on this screen give you second looks at items that we've already discussed. The Audio button brings you to the dialog boxes shown in Figs. 10.3 and 10.4. The Video button brings you to the dialog box shown in Figure 10.7. It's always helpful to review all these options just one more time before starting the capture.

The Compress button brings up the dialog box shown in Fig. 10.13. This dialog box is available for use by the various capture boards, but few use it. You can typically ignore this box during normal operation.

Figure 10.13 Video Compression dialog box

9. Capture

MANUAL CAPTURE

At this point, if your analog source isn't controlled through an MCI or VISCA controller, you're ready to go. Press OK in Fig. 10.12 to trigger the dialog box shown in Fig. 10.14.

Now it's up to you to press OK at the right time. The video should be showing in the window underneath the dialog box in Fig. 10.14. It's obviously easier to watch the video if you move the box away. When the target sequence begins, click OK.

Although the system looks ready to capture, in reality the board isn't quite ready, and will need between two and 10 seconds to start the capture after you click OK. This means that you really have to press OK between two and 10 seconds *before* your target video appears, which takes some getting used to. However, you can easily crop extra frames after capture, so it's usually faster to press OK early and capture too much than to try and time it right and miss the first couple of frames.

Once you press OK, a message appears in VidCap's Status Bar advising you to press "Escape" to end the capture. During capture, depending upon the capture board and capture format, the video in the screen either will or won't update. For example, when capturing with the Smart Video Recorder in

Figure 10.14 Launching a non-MCI controlled capture

Figure 10.15 Capture results presented in VidCap's Status Bar

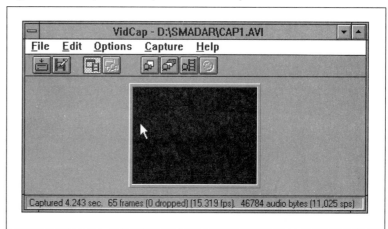

Indeo 3.2 mode, sometimes the screen updates and sometimes it doesn't. In raw mode, it doesn't update the screen at all and stays fixed on the first frame until you end the capture. Either way, VidCap presents a running frame count and capture time, updated after each 100-frame interval, in the Status Bar (see Fig. 10.15 for location of Status Bar).

This lack of visual feedback makes an external NTSC monitor very helpful. Alternatively, you can time the clip or count the number of frames and use the capture information presented by VidCap to stop when appropriate.

After you've completed the capture, VidCap presents capture results, including duration, number for video frames, frames dropped and audio bytes captured in the Status Bar. You've undoubtedly captured extra frames of the beginning and end of the video. You can load VidEdit by pressing the VidEdit icon in VidCap's toolbar. Next chapter we'll describe how to use VidEdit to carve away these extra frames.

Right after capture is a good time to either change the name of the capture file so you don't write over this file during the next capture, or use the Save As command to save it to a new name.

COMPUTER-CONTROLLED CAPTURE

As we discussed, step capture requires an analog source that's both frame-accurate and capable of being controlled by your computer. However, if your deck is computer-controllable and not frame-accurate, you can still use it to drive real-time captures (Fig. 10.16). This avoids the hassle of pressing "OK" at just the right time, and minimizes post-capture editing. More important, if you record time code or frame settings, you'll be in great shape if you have to capture the same footage again.

COMPUTERIZING YOUR DECK

Let's look at how to "computerize" two devices that are both frame-accurate and computer-controlled, the Pioneer CLD-V2600 laserdisc and the VISCA-controlled Sony CVD-1000 Hi-8 deck. The procedure is roughly the same for all computer-controlled decks, whether frame-accurate or not, so this should help those looking to perform real-time capture as well as step-frame.

Before getting started, let's overview how and where Windows keeps track of multimedia devices, including the MCI-

Figure 10.16 MCI-controlled video capture

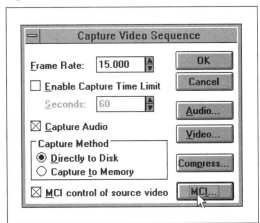

controlled devices discussed here and the codecs discussed in Chapter 12.

INSTALLING STUFF—THE HARD WAY

For the most part, multimedia products contain setup files that load the necessary drivers into the necessary places and create nice little icons you can click to load the program. Life on the bleeding edge is a bit less certain. What happens, for example, when you download the Pioneer driver from CompuServe and find that you have to load it yourself?

If the thought of messing with your System.ini file makes you break out in a cold sweat, you might consider skipping ahead. On the other hand, I guarantee there will come a day when you try to load a driver and your techno-buddy is unavailable and you're faced with the "learn something or punt" dilemma.

So read on. It's not rocket science and you're only a couple of mouse clicks away from being a more robust Windows user.

SYSEDIT

Every Windows installation has a System.ini file that tells Windows which drivers are loaded. Every driver that works with Video for Windows or the MCI specifications has to register in System.ini. This is the file, for example, that VidCap will check to see if a certain driver is loaded, the name of the driver and its location.

Fortunately, Windows includes an easy mechanism to review your Windows.ini file. It's called Sysedit, and if it's installed, it's usually located in your Main Windows group. If not, click on Main, and then from Program Manager, select File, New, and Program Item. The screen shown in Fig. 10.17 will appear.

Type in "c:\windows\system\sysedit.exe" in the Command Line box and complete the other information shown. Press "OK" and the icon shown in Fig. 10.18 should appear.

Double-click on the Sysedit Icon and you'll be rewarded with the screen shown in Fig. 10.19.

Figure 10.17 Loading Sysedit in the Program Item Properties dialog box

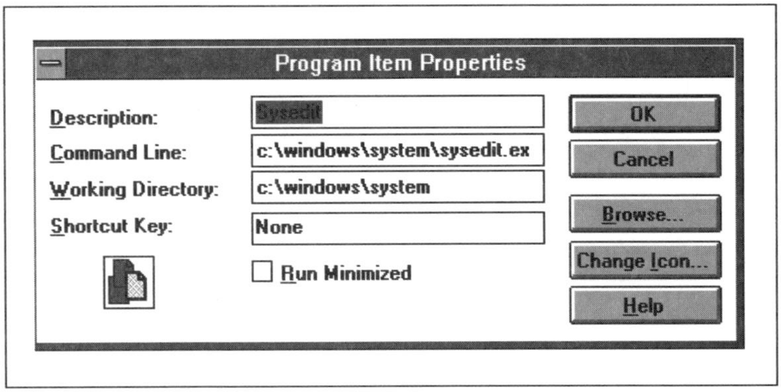

These four files, Autoexec.bat, Config.sys, Win.ini and System.ini, contain the basic DOS and Windows startup information for your computer. These are the files typically effected by product installations; these are the files that activate and often fail to activate your multimedia drivers.

By touching any of the file windows, you can edit the file. Let's take a closer look at the System.ini file, where we have to

Figure 10.18 SysEdit—the window to all your configuration files

Figure 10.19 The System Configuration Editor

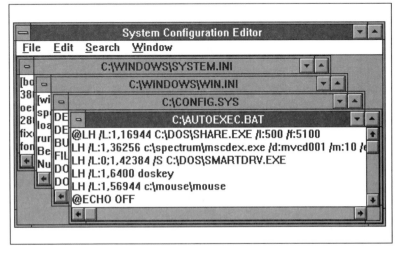

Figure 10.20 Your System.ini file

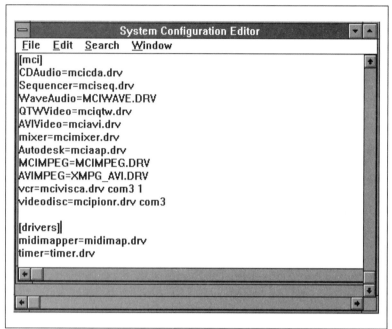

install the Pioneer and VISCA drivers. As you can see in Fig. 10.20, instructions are grouped in under headings identified with brackets. In this chapter we'll focus on the [mci] section.

The first few times you work with your System.ini file directly, make a backup copy as we discussed in Chapter 9. It's always easier to start again than to try and recreate your changes without a backup.

THE DRIVERS

To computerize your analog deck, you need (a) drivers that are (b) installed in the correct location, and (c) the correct physical connection. We'll start with the drivers.

Table 10.7 shows the drivers that you'll need for the Pioneer laserdisc and Sony CVD-1000 Hi-8 deck, and how to get them. **NOTE THAT BOTH SONY DRIVERS ARE NEEDED FOR THE VISCA CONTROL.**

Table 10.7 Drivers to run computer-controlled analog decks and where to get them

Device	Driver	Size	Date	Where to Get
Pioneer laserdisc	mcipionr.drv	60848 KB	10/28/92	Video for Windows 1.1 SDK/Video for Windows 1.0
Sony VISCA	mcivisca.drv	96912 KB	9/23/93	Sony bulletin board 408-955-5107
Sony VISCA	vcr.mci	4744 KB	11/19/93	Video for Windows 1.1 SDK

Once you have the drivers, copy them into your Windows\ system directory. Then, to let Windows know where they are, modify your System.ini file as shown on the last two lines of the [mci] section of Fig. 10.20.

These two lines break down as shown in Table 10.8.

VCR=mcivisca.drv
Videodisc=mcipionr.drv
This identifies the drivers for the particular device. The Sony uses mcivisca.drv and vcr.mci, and the Pioneer laserdisc uses mcipionr.drv. If you're installing these devices, copy this portion of the line exactly.

com 3
This line tells your Windows which communication port, or com port, the analog deck connects to. Com ports are serial interfaces to and from your computer. Computers come with four, although com1 and com3 share an interrupt, as do com2 and com4. This means that only two ports can be active at any one time.

Com ports run either through boards installed in your computer, like modems, or through a serial port. Most computers have two serial ports, one a nine-pin connector where the mouse typically sits,

Table 10.8 System command lines for the Sony CVD-1000 and Pioneer CLD-V2600

Device	Windows.ini line
Sony CVD-1000	vcr=mcivisca.drv com3 1
Pioneer CLD-V2600	videodisc=mcipionr.drv com3

the other a twenty-five pin connector. You'll need a free serial port for a cable connecting your computer and the analog deck.

If you don't know which com port is open, go back to the Microsoft Systems Diagnostics (MSD) we discussed earlier when charting interrupts. Find your mouse and your modem. Since you'll need your mouse during capture, you'll ordinarily have to configure the analog deck on the same interrupt as your modem.

For example, if your mouse is on com2, don't configure your analog device on com2 or com4. Use com1 or com3, whichever isn't occupied by your modem. At a minimum, this means you can't capture and use your modem simultaneously. At worst, your modem may not work at all, which is what happened on our Gateway.

1 (Sony)   The VISCA control enables multiple devices. This tells Windows that this device is the first one. To install a second device, you would change the command line in System.ini to the following command:

VCR1=mcivisca com3 2

Windows checks the System.ini file when it loads, so changes won't take effect until you exit and restart Windows. Check the Drivers dialog box in the Control Panel (Figs. 9.1 and 9.2) and you should see the driver loaded as shown in Fig. 10.21. The VISCA driver has a similar setup listing the com port selection and number of VCRs configured.

If you don't find the driver in the control panel, it usually relates to a syntax error—Windows won't check your serial port until you actually try to load the driver in VidCap. Review your entries in System.ini and make sure the drivers are in the Windows\System directory. This should fix the problem.

Figure 10.21 A properly configured Pioneer driver in the Drivers section in Control Panel

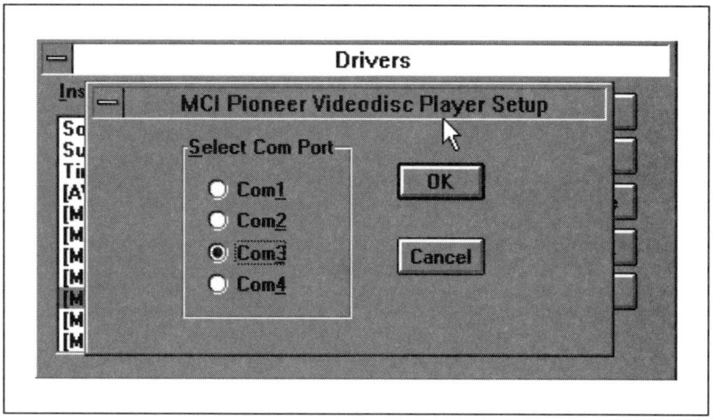

Drivers

MCI Pioneer Videodisc Player Setup

Select Com Port

○ Com1
○ Com2
⊙ Com3
○ Com4

OK

Cancel

CABLES AND CONNECTORS

Both devices have specialized cables. The Sony CVD-1000 ships with the VISCA cable shown on the right side of the deck in Fig. 10.22. The cable has an eight-pin Min-Din connector that connects directly into Macintosh computers, but not IBM

Figure 10.22 The Sony CVD-1000, or VDeck, and associated peripherals

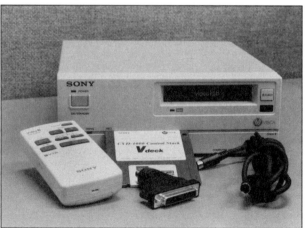

computers—you'll need the adapter shown lying on the disk in front of the deck.

Which adapter depends upon the serial connector that you'll devote to this connection. There are three kinds of serial port connectors, nine-pin male (pins sticking out) or DB9 (male), and 25-pin (DB25) male and female connectors. You'll need the opposite type of connector to the one open on your serial port. For example, if your target port is a DB25-pin male, you'll need a female connector. See birds and bees for details.

The other side of the connector must be the DIN8 connector to hook to the VISCA control. Any of the three connectors should cost under $10.00. You will need this connector if running on an IBM computer, so you should be sure to pick one up when and where you purchase your deck.

Like the cabling, the disk in the picture contains only Mac software—I guess Sony doesn't think the DOS/Windows standard will catch on. Anyway, you'll have to upload their drivers from their bulletin board to get started.

The Pioneer cable, not shown in Fig. 10.23, costs about $20.00. A normal serial cable won't work—the pinouts on the laserdisc end are nonstandard. The cable typically isn't found in consumer stores such as Radio Shack, so if you didn't get one when you purchased your laserdisc, you'll have to track one down in your local A/V store.

Figure 10.23 The Pioneer laserdisc, desktop video quality champ, and remote

You'll also need the video and audio cables. Both decks use S-Video cables, the four-pin connector shown in Fig. 7.1. (the middle rectangle is a guide, not a pin; see p. 132.) The cable has the same connector at both ends, as do the one-pin composite cables.

The CVD-1000 has two main rows of connectors on the back, which look like the drawing in Fig. 10.24. The input side takes input from another deck or computer-generated analog output. We'll ignore these. The output side contains two rows, with one S-Video output, two composite video outputs, and two audio outputs. When using this deck, connect the S-Video output directly into the S-Video connector on your capture card.

You can use either audio output. Typically, you'll have to use a y-cable that combines the left (white) and right (red) audio feeds into one jack for input into your sound card. Most sound cards include at least one such cable. Incidentally, most decks and cables are color-coded, which really helps makes the task a lot easier.

The monitor out outputs go to your NTSC tracking monitor and optional speaker. Most higher-end audio cards will output

Figure 10.24 The business end of the Sony CVD-1000, showing analog inputs and outputs and ports for VISCA controls

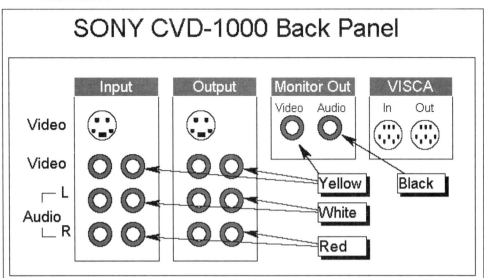

audio simultaneously during capture, so you won't really need the separate speaker. One extra benefit of using the NTSC monitor with the Sony deck is that the monitor out shows time code information, which is essential to identifying start and stop points on the tape. This helps make up for the fact that the Sony doesn't have an LCD readout of the time code information like the laserdisc and most higher-end decks.

Without an NTSC monitor, you have to hook the monitor out feed into your capture card to track time codes, which means switching back and forth between the composite and S-Video feeds. That can be a real pain administratively, and, as we saw last chapter, the composite signal can introduce noise into the S-Video signal during capture. Best advice is—if you use the Sony, get an NTSC monitor.

While all tape decks will have slightly different configurations, most professional decks and even pro-sumers will have inputs, outputs and monitor outs. So before you plug into your deck, find the input and output jacks and make sure you use the output.

The laserdisc will be even simpler because it doesn't have video and audio inputs, since you can't record on the finished disk. You could confuse the composite video out with the audio outs, but most are pretty well labeled, so if you pay attention you'll have no problem.

Let's summarize our equipment needs:

1. S-Video cable, which typically comes with cameras and professional decks, but not consumer decks such as the laserdisc.
2. Audio cable, two-pin to one-pin, which typically comes with your sound card.
3. Cable hookup between computer and deck. In the case of the Sony, you'll need the VISCA cable—which comes with the VDECK (**remember the adapter**). If using the laserdisc, you'll need to pick up the optional Pioneer cable.
4. Composite cable to NTSC monitor—this is typically supplied on consumer decks but not professional.
5. Audio cable to audio out (optional).

Figure 10.25 MCI capture command screen

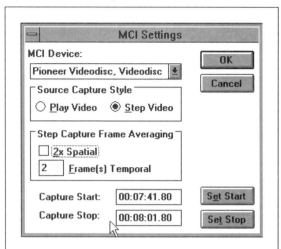

STILL-FRAME CAPTURE

Press the MCI button shown in Fig. 10.16 to bring up the dialog box shown in Fig. 10.25. Before loading this dialog box, Windows checks to see which device is attached to the serial port and loads the proper driver, in this instance the laserdisc driver.

If no driver loads in the MCI device box, your installation isn't correct. If you haven't already checked if the driver is loaded in the Drivers dialog box in the Control Panel (Fig. 10.21), do it now. If it is loaded, either you've selected the wrong serial port, or the link between the computer and the analog deck is defective.

If you select Play Video, you'll have real-time, computer-controlled capture. Select Step Video to go the step-frame route.

STEP-FRAME CAPTURE AVERAGING

Here's what VidCap has to say about the Step Capture Frame Averaging controls:

The Step Frame Averaging: The following two options are used with the Step Video capture method to bring out detail and reduce video noise:

2x Spatial specifies whether to expand the frame size during capture to bring out more detail in the image. VidCap doubles the frame size captured and then uses spatial averaging to reduce aliasing effects. This technique enhances fine lines in the sequence, at the cost of some fuzziness.

n Frames Temporal specifies whether to reduce video noise by repeatedly capturing frames and averaging out the noise from the captured images. Specify how many images to capture and average per frame. Note: Some capture boards provide hardware routines for performing these enhancements. If your capture board includes such a feature, you should not select any of the step-frame averaging options provided by VidCap.

As shown in Figure 10.5, the interpolation and filtering done by most hardware cards is superior to that performed by Video for Windows. We tested 2x spatial averaging and found virtually no difference between the nonaveraged and 2x averaged.

This could be, as Microsoft says, because our capture card is zooming anyway. You might test for yourself, but in the absence of proof that it does improve your video, I'd recommend against 2x spatial averaging.

The N frame averaging is a more interesting question. In concept, this is a great idea. Remember that during digitization, your capture board has to assign each pixel one color out of 16.7 million possible colors. The sheer number and closeness of choices virtually ensures that the captured value will differ from the real value to some degree. N frame averaging lets you digitize the same frame a number of times and average the pixel values to derive one averaged value, which should more accurately represent the real color.

When you select three, for example, the capture board digitizes each frame three times and averages the values. In theory, this should help interframe compression, since there would be more interframe redundancy, and intraframe compression, which would benefit from internal redundancy. Alas, a

Table 10.9 Frame averaging test files

Video Type	Averaged File	Non-Averaged
Low motion-Indeo 3.1	tlkstep3.avi (2,612,422 bytes)	tlkstep1.avi (2,607,548 bytes)
High motion-Indeo 3.2	actstep3.avi (3,019,716 bytes)	actstep1.avi (2,953,666 bytes)

panacea it's not, and in application we saw very little difference between files that were averaged upon capture and those that weren't.

We tested two kinds of files, high- and low-motion sequences (see Table 10.9). The control file was captured without averaging, and the test file was captured with three frame averaging. In the low-motion videos, there was no perceptible difference. In the high-motion sequences, when there was a difference, it typically manifested itself as artifact in the averaged file that wasn't in the non-averaged file. Translation: averaging produced artifacts in laboratory rats . . . er, videos.

On the other hand, I recently received a facsimile from a product manager of a leading capture board manufacturer stating that they found that 3x averaging "cleaned the video up" and "marginally improved visual integrity and helped compression." As you can see from Table 10.9, we found that the averaging *increased* the average compressed frame size of the action file by 197 bytes, and the size of the talking head file by 11 bytes. The difference isn't statistically significant, but certainly counterintuitive of the product manager's claim.

Once again, this may relate to your capture board, so you may want to try one or two short files yourself to see if you can spot a difference. For the Smart Video Recorder Pro, we recommend against averaging—it caused artifacts in some videos and increased the file size in all of our tests.

CAPTURE START AND STOP

Laserdisc—When working with the laserdisc, life gets really simple. Use the remote control unit to move the laserdisc reader to the desired start frame and then press the "Set Start" button. VidCap checks the location of the reader and enters the proper time code. Then move the reader to the desired stop frame, and press the "Set Stop" button to enter the proper stop time. Dutifully record either the frame numbers or time code information in your trusty notebook and you're ready to capture. Press OK to return to the Capture Video Sequence Screen shown in Fig. 10.16.

If your laserdisc has an LCD counter, it's pretty simple to track your start and stop points. If not, you'll have to use the display control to show the frame number in the screen itself. Remember to shut off the display control before you start to capture because what you see is what you get, and if the frame number is visible, it will appear in your video.

Remember also that laserdisc step frame capture requires a CAV disk. Most commercial laserdiscs are *not* CAV, so if you're capturing from a commercial laserdisc and having problems, it's probably the format.

Sony CVD-1000—Setting up the Sony is a bit more difficult. The V-Deck is frame-accurate, but there are no step controls that can step you to a precise frame. The easiest way to capture is to estimate time code information by watching the tape on your NTSC monitor. Enter in the values where indicated and press OK to return to the Capture Video Sequence Screen shown in Figure 10.25.

If you don't have an NTSC monitor, you'll have to connect both the composite and S-Video cables to your capture board and toggle back and forth between the two signals. Your composite signal will have the time code information and the S-Video signal will be your capture signal. It's not pretty, but it works.

One significant hassle relates to the VISCA control itself. When the VISCA cable is connected and the driver initialized, the deck is under VISCA control, and all other controls are frozen. Exiting the dialog box didn't release the VISCA control.

This means that you can't use manual controls to find start and stop points. Since VidCap is a dumb driver designed to capture, not hunt and seek, there's no software to help you seek either. The manual states that VISCA shuts down when you power down the unit, but this wasn't my experience.

Which leaves two options. You can find all the start and stop points on the tape and then start capturing, which sounds terribly responsible and almost "Skinner-ish" in that it totally delays gratification, or pull the VISCA cable after every capture, manually find the start and stop points and then plug the driver back in to capture. Guess which I did?

Sony lists the following points to consider when working with their VISCA drivers in the Install.txt file included with their driver.

1. Be sure to reset your camcorder hms counter at beginning of the tape; you will not be able to search into negative counter values (either externally or with the mci-string "set vcr counter 0").

2. Time code may not be detectable, even though it is present, on some decks at the very beginning or the very end of a tape. If you are using mplayer, play the vcr for a few seconds and then close and reopen the driver, or issue the "time mode detect" command.

3. Index marking is highly hardware specific. Please refer to your video deck/cameras operation manual.

4. The millisecond time format assumes 30fps (as opposed to 29.97). This is done for compatibility reasons. SMPTE time formats use the native time format on the tape, and should be used whenever possible to avoid rounding errors.

REAL-TIME CAPTURE CONSIDERATIONS
With either device, when capturing in real time, some boards need a few frames to get rolling and actually start to capture, even when the deck is controlled by the computer. Get in the habit of adding about half a second before the first frame you actually want to capture. In Fig. 10.25 (p. 249), for example, if we were capturing in real time we would have started at 7:41:30 instead of 7:41:80. When capturing in real time, be sure to check each captured file to make sure that the capture started in time and captured all the frames that you needed. When capturing in step-frame mode, you can specify the exact

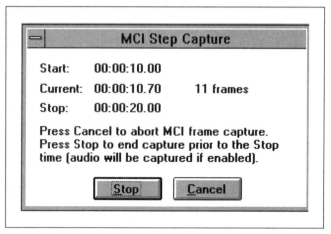

Figure 10.26 Our reward—a step capture in action

start and stop frame because the capture board starts immediately on the first frame.

CAPTURE AT LAST

Press OK in the Capture Video Sequence screen (Fig. 10.25) and you'll be rewarded with the screen shown in Fig. 10.26—if you're step-capturing.

Now it's time to sit and wait for the capture to finish, which takes about a frame a second. If you're capturing in real time, the dialog box shown in Fig. 10.14 comes up, prompting you to press OK, which will start the capture.

SUMMARY

1. Your primary target system controls capture parameters such as resolution and frame rate. Always identify one configuration (e.g. MPC-2) and benchmark your video on that system. Then consider the trade-offs inherrent in that decision relating to higher- and lower-end computers.

2. **Video Resolution**—Here are our views by video type.

(a) *Talking Head*—prioritize synchronization over resolution. Users prefer smaller screen, better-synchronized video over larger, less synchronized video. 240x180 resolution is fairly optimal.

(b) *Moderate Motion*—Prioritize resolution and image clarity over frame rate. Try 10 fps for these videos at 320x240 resolution.

(c) *High Motion*—Prioritize smoothness (e.g., display rate) over resolution. However, since Cinepak is so fast, you may be able to get away with 320x240 and get both.

Always capture at your target resolution. Scaling after capture can distort the results.

3. **Capture Formats**

(a) *Step Frame*—Always capture raw.

(b) *Real Time*—Capture compressed at the highest possible data rate that still gets all frames to disk.

4. **Video Source**—Favor S-Video over Composite.

5. **Frame Rate**

(a) Use 15 unless you'll be working exclusively with a video co-processing card such as VideoLogic's Movie928. Consider 10 fps when working with moderate-motion sequences where content is primarily in the video (e.g., repair and other "hands-on" videos). When working with a video co-processor, try capturing at 30 fps in lower resolutions (e.g., 320x240) and using the card to scale the video.

(b) Always capture at the target frame rate. Modifying frame rate after capture can distort the results.

6. To "computerize" your capture system, you need (a) drivers, (b) properly installed, and (c) the required physical links or cables. Capturing from a computer-controlled deck is preferable whether you're step-framing or capturing in real time.

7. Step-Frame Capture Averaging produced no tangible benefits in any of our tests, and we recommend against it.

PREPROCESSING
FOR
COMPRESSION

11

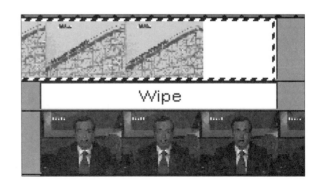

PREPROCESSING FOR COMPRESSION

We've captured our video. Hopefully in raw format, but perhaps lightly compressed during capture.

Before compression are three preprocessing tasks: clipping to the final video size, filtering, and transitions and other effects. We'll perform these first, and then compress to final form to minimize the number of times we compress the video.

We'll perform nips and tucks with VidEdit, the easiest tool to use for small video editing jobs. Typically this involves shaving extra frames from the beginning and end of the video.

For certain low-motion videos, filtering produces higher-quality video at lower bandwidths, an overall slimming effect that's quite lovely. We'll look at the tool that makes this happen and when and how to use it.

Finally, we'll go "stylin' " with video editing programs that let you add transitions, titling and special effects to your video. These bring you one step closer to Hollywood, but bring a unique set of system requirements and their own brand of compression concerns.

Nips and Tucks with VidEdit

If you're working with straight, linear video without transitions, clipping is fairly straightforward. VidEdit lets you work with icons or commands (Fig. 11.1). If you're an icon kind of guy or girl, note the scissors in the upper left hand of the toolbar. This cuts frames.

Figure 11.1 Note scissors up top, Audio/Video selectors in lower right-hand side and Mark In and Mark Out icons

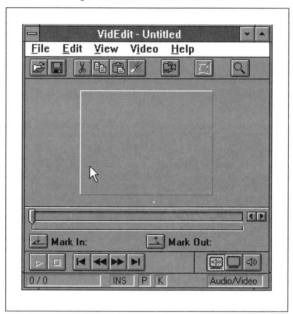

Note the three icons in the lower right-hand side of VidEdit, consisting of a television, a horn and a television with the horn inside. These select the tracks affected by your editing decisions. If the horn is pressed, for example, and you cut a frame, only the audio is cut. The empty television cuts only video frames, and the remaining icon cuts both audio and video.

Note the slider bar below the video and the Mark In and Mark Out controls. These select the frames affected by your editing decisions. When you mark a frame in, it *will* be cut. In contrast, the frame that you mark out will *not* be cut. So, when using the slider bar, start with the frame you want cut, and end with the frame after the last one you want cut.

To trim the unwanted frames from the start of the video, press Mark in on the first frame (frame 0) and use the single-frame controls to the right of the slider bar until the first frame you want to keep is onscreen. Press Mark Out. Make sure the correct track is selected, and press the scissors.

To trim unwanted frames from the end of the video, page over until the first frame you *don't want in the video* is on screen. Select Mark In and then page over until the final frame of the video. Select Mark Out, check the track and punch the scissors. Mission accomplished.

Insert vs. Overwrite Mode

As you can see from the lower VidEdit toolbar in Fig. 11.1, we're currently in Insert Mode (the INS letters in the second box from the left), which we covered back in Chapter 4. A quick review here will probably help.

VidEdit has two modes, Insert and Overwrite, which you select in the Preferences dialog box under Edit (see Fig. 11.2). When you cut frames in Insert mode, the frames and their frame slots both go away. For example, if you started with 300 frames and cut 15, you would also cut the frame count down to 285. Contrast that with Overwrite Mode, which deletes the

Figure 11.2 Preferences screen, located under Edit, contains the toggle switch for the Insert and Overwrite Modes

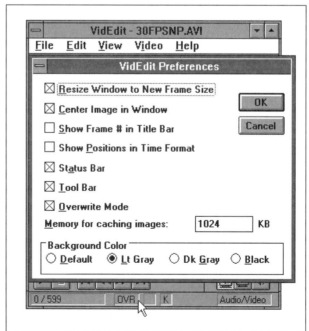

frames but saves the slots. If you cut 15 frames from 300 in Overwrite Mode, you'd still have 300 frames, but the first 15 would be blank. At some point in the future, if you find yourself cutting frames and getting frustrated because they won't go away, check your mode. Obviously, use Insert Mode when nipping and tucking with VidEdit.

Those who feel more comfortable working with menu commands can find controls for frame and track selection and cut and paste under the Edit line.

Where to Stop and Start

Now is the time to plan for transitions into and out of your videos. For example, the primary test video that you've seen throughout the book transitions from black to video in about one second. We did this by moving to the actual start frame and paging back 15 frames. Ditto on the back end, where we started at the last video frame and counted 15 frames forward to handle the fade to black.

Figure 11.3 is a screen shot from Asymetrix's new Digital Video Producer showing a transition from an action shot to a

Figure 11.3 Screen shot from Asymetrix's new Digital Video Producer, showing transition between roller coaster scene and spokesman

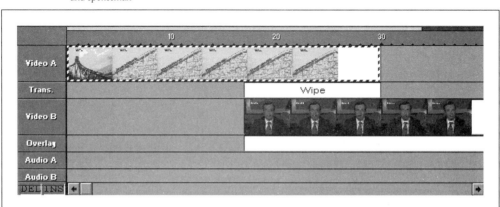

talking head. While the talking head footage needs to be available for the entire transition, which takes about one second, he shouldn't start talking until the transition is complete. If you don't factor this into the nipping and cutting process, you'll be in trouble when it times to assemble the final videos. Finally, after trimming unwanted frames, remember to select the "no recompression" option when saving the file.

SLIMMING WITH DOCEO FILTER

Throughout this book we've talked about the imprecision of analog tape formats and the inherent inaccuracy of the digitization process. Capture boards select one color out of 16.7 million, pixel by pixel, frame after frame, and are extremely unlikely to select the same value for two consecutive frames, even if the analog footage hadn't changed at all. This was the problem that Microsoft sought to attack with their 2x frame averaging during step capture (see Table 10.9, p. 251.)

As we learned in Chapter 2, motion is the enemy of interframe compression, the real power behind most codecs. If a value for the same pixel on consecutive frames changes by even one color, the codec sees motion, and must devote valuable video bandwidth to comprehend the change. This takes bandwidth away from real motion, which suffers in quality, and causes the background shimmer that's become characteristic of digital video.

This is the problem we've addressed with the Doceo Filter. Through a sophisticated but simple-to-use process, the Filter smoothes interframe noise improving compression efficiency and video quality. While it's not perfect, and works best for specific video types, it is extremely effective in these circumstances. Let's see how it works.

It's difficult to tell from the flat picture on a page, but Fig. 11.4 is noisy video. We've adjusted Fig11_4l.avi to help illustrate this point. It's a raw file, which we've slowed to one frame per second to let you see the noise in the walls, on the desk, pretty much throughout the video. If you're following along, after playing Fig4_11l.avi on the left, load Fig11_4r.avi on the right and play the file, focusing on the areas that look

Figure 11.4 Noisy analog video—the bane of compression

noisy in the first video. You'll see that much of the noise is gone. That's the Doceo Filter at work.

How it Works

The Filter analyzes the capture file and creates a new filtered file, leaving the original untouched. You load the capture file, input several parameters and press start (see Fig. 11.5). The program reviews your input, analyzes the video file and computes the optimal filter parameters.

The program loses the ability to distinguish between noise and true motion as the amount of true motion increases. For example, it's pretty tough to distinguish between noise and motion in an action sequence where every pixel is moving in every frame.

On the Video Compression Sampler, we filtered two sequences, the talking head and meeting clips. We did not filter the install or high-motion clips. Our tests revealed quality and compression improvements for Video 1, Indeo 3.1 (Quick and SuperCompressor), Indeo 3.2, Cinepak and RLE. Since Xing uses intraframe compression, filtering had little effect on these files.

Figure 11.5 Control Panel for Doceo Filter

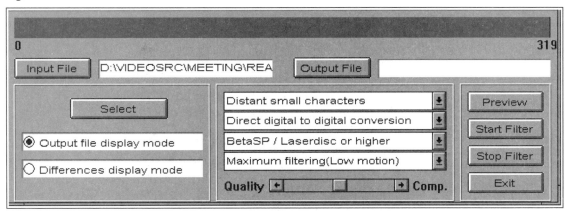

As we'll see, the Filter has two functions to help expand its use. Preview Mode lets you preview exactly what the filter considers noise and what it considers real motion. If it's not accurately assessing the video, you shouldn't use it for that sequence. The Video Selection screen lets you filter only static segments of the video, and exclude motion sequences such as transitions. This lets you filter talking head shots and helps expand its use dramatically.

Filter Parameters

The Filter Parameters section lets you enter details about the amount of motion in your video, the quality of your analog sources and your capture technique. This lets us assess how to filter the video. The quality vs. file size slider bar allows you to further customize the settings. Before starting, the Filter also analyzes the video itself to determine the final filter parameters.

This rating technique assumes a video that is relatively stable (e.g., shot from a tripod) with all pans, zooms and cuts excluded from the filtering process with the Video Selection control. Table 11.1 gives the breakdown in values from the product manual.

The content description in Table 11.1 seeks to determine the nature of the video motion. Small subtle movements by distant characters get obscured through filtering. On the other hand, if the screen is largely composed of one individual or object, the movements by nature tend to be rather gross, covering a number pixels.

Capture method (Table 11.2) in large part determines the extent of interframe noise. The footage from Horizons Technology, for example, had little noticeable interframe noise, so heavy filtering could damage, not enhance, the ultimate video. Real-time capture with ISA bus capture cards creates the most interframe noise because of the compression necessary to get the frames to disk. This necessitates a higher compression value.

Analog source (Table 11.3) obviously relates directly back to our Chapter 7 findings. Color fidelity and signal bandwidth

Table 11.1 Values for Content Description Selection

Value	Content Description
1	Two or more characters filmed at a relative distance (e.g., meeting video).
2	Two or more characters in video, but still taking up substantial portions of the video frame.
3	One major character or object dominating the video frame (e.g., talking head video).

Table 11.2 Values for capture method

Value	Capture Method
1	Direct digital-to-digital conversion (a la HTI)
2	Step-frame or other raw capture
3	Real-time at data rates exceeding 500 kB/second (320x240@15 fps)
4	Real time at data rates below 500 kB/second (320x240@15 fps)

Table 11.3 Values for analog source

Value	Analog Source
1	BetaSP or higher analog format; laserdisc.
2	S-Video or Hi-8
3	VHS or 8 mm.

Table 11.4 Filtering options

Value	Filtering Method
1	Maximum filtering (low motion)
2	Medium filtering (low to moderate motion)
3	Artifact avoidance (moderate to high action)

lessen signal noise, and which must be considered in the filter parameters.

Table 11.4 illustrates the three filtering options available for the Doceo Filter. The first, maximum filtering, should only be used with extremely low motion, such as talking head videos with low amounts of background noise. The second, medium filtering, should be used with videos containing low to moderate amounts of motion or noise. The third option, artifact avoidance, is designed for moderate- to high-motion videos. While it may only produce marginal benefit, it will tend not to create artifacts that the other filtering methods might on these video types.

Figure 11.6 The Quality slider bar

Figure 11.7 The two ends of the quality/compression spectrum. The picture on the left is the original, the middle picture filtered at maximum compression and the picture on the far right filtered at maximum quality

The best way to determine which setting to use is to experiment with the Preview mode in Differences display mode (Fig. 11.7). Try the different methods and see which does the best job of filtering out background noise while retaining the motion essential to the video content. Once you pick the best method for a particular type of video footage, you probably can use that value for all videos of that type.

Quality Slider

The Filter program settings are designed to produce the optimum blend of video quality and low bandwidth. The quality slider bar lets users optimize compression over quality, or the reverse, depending upon their goals and requirements.

This is shown in Fig. 11.7, which also illustrates the filter's Preview Mode. On the left is the original video. The middle video shows the filtering at maximum compression, the video on the far right shows maximum quality.

The white spaces in the two filtered pictures are areas that are considered "noise" and are filtered. The other areas are real motion areas that aren't filtered. The middle video will yield the highest compression because most of the picture is filtered. However, small details of the face will be lost and the video may look washed out.

The picture on the right will yield the highest video quality because the fine details around the face are preserved. As you can see, however, much more noise from the back wall and coat also seeps through.

Noisy Video

The next three figures compare filtered and unfiltered frames extracted from a video captured in real time from a laserdisc, at a data rate of about 370 kB/second. Except for the laserdisc source, this is pretty much a worst-case analysis.

Figure 11.8 shows the videos compressed with Indeo 3.1— at a quality setting of 100. Rather than compress both files to 150 kB/second and compare quality setting, we decided to measure the pure data rate differential, which turned out to be quite substantial. The filtered file compressed to 161 kB/second, while the nonfiltered file weighed in at 228 kB/second, fully 67 kB/second larger. Subjectively, the filtered file looks

Figure 11.8 Filtered vs. unfiltered comparison, Indeo 3.1

Figure 11.9 Filtered vs. unfiltered comparison, Indeo 3.2

substantially clearer, notwithstanding the significant file size differential.

Figure 11.9 compares the same videos compressed with Indeo 3.2. We compressed identically, of course, to a target data rate of 150 kB/second. The filtered video came through at 105 kB/second, while the unfiltered file squeaked through at 150 kB/second. Even at a bandwidth almost 50 kB/second lower, the filtered video looked much better than the unfiltered, with much less background motion.

Indeo 3.2 fans will be quick to point out that the filtered Indeo 3.2 video is a touch clearer than the Indeo 3.1 Quick Compressor. Es verdad! The biggest problem with Indeo 3.2 in low-motion videos isn't the still-image picture quality, which is absolutely outstanding; it's the noise in the background. Compare the two videos (Fig11_8l.avi and Fig11_9l.avi) side by side and you'll immediately see why we favor the Quick Compressor.

Figure 11.10 Filtered vs. unfiltered comparison, Cinepak

Finally, Fig. 11.10 compares the same two videos compressed with Cinepak. While we don't recommend using Cinepak for these kinds of videos, for obvious reasons, filtering did improve video detail by a small degree. Cinepak was the only codec where the filtered clip was larger than the unfiltered, with the videos weighing in at 150 kB/second and 140 kB/second, respectively.

Other Types of Videos

This test video was obviously a worst-case analysis. However, even the Holy Grail, the direct digital-to-digital video captured by Horizons Technology, couldn't reach 150 kB/second with Indeo 3.1 at a quality setting of 100—we had to filter. In fact, pretty much every video displayed in the book was filtered, which contributed greatly to overall video quality.

Figure 11.11 Spots created by filtering through a fade from black opening transition

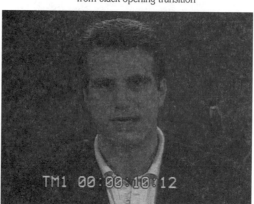

Filter Faux Pas

During the early stages of working with the Filter we discovered an aversion to fade-in and fade-out transitions. In fact, as you can see from Fig. 11.11, the filter created some pretty "spotty" results. This finding led to the filter selection screen that lets you exclude frames like these that would be adversely affected by the filtering. This screen is shown in Fig. 11.12.

Figure 11.12 Filter Selection screen

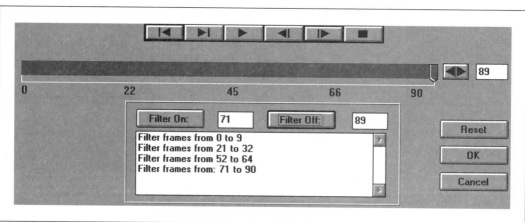

Enter the Filter Select screen by pressing "select" on the main screen. The Filter assumes that the filter is "off." Using the controls provided, you simply step through the video frame by frame turning the filter on and off as needed. You can use Preview Mode (Fig. 11.7) to hunt for similar artifacts in your video, and then exclude these frames.

STYLIN' WITH VIDEO EDITORS

Video editing programs bring a lot of fun to what otherwise can be the drudgery of video preparation. You can also polish your videos to a degree not possible with simple tools such as VidEdit.

The Compression Black Box

Space and time don't allow us to exhaustively describe what these programs can do, so we'll focus on some compression-centric issues. The first is pure compression. In addition to editing, these programs will replace VidEdit as your primary compression interface. The essential compression values such as key frame and quality setting will still be the same, they'll just be in different places. However, will the compressed file be the same?

More than almost any other area of Video for Windows, the VidEdit compression functions have been a real black box that Microsoft has shielded from all other developers.

This is a fairly long-winded way of saying that the compressed files created by the video editor may not be the same as those created by VidEdit (they use the same codec but often don't work the same as VidEdit). We know from developing the Doceo batch compression program that matching VidEdit's performance for all the codecs on every file is extremely difficult. Early versions of Adobe Premiere suffered from this problem, and I'm certain that early versions of most other programs will stumble as well.

When compressing with a video editor, be sure to test the compressed file against the same file compressed with VidEdit, especially if you're developing for CD-ROM applications. We had an extremely difficult time matching VidEdit when data-rate management and CD-ROM padding optimization issues arose. Smaller-resolution files that compressed to well under the target data rate were usually not a problem. You can check your file against VidEdit by outputting the file from your video editor in raw format and compressing with VidEdit.

Load both files into VCS Play and check the display rate. Some of our early files played back much more slowly than VidEdit's files. Then check the frame profile, all through the file. An exact match isn't necessary and is unlikely with some codecs. However, in bandwidth-limited applications, check that the target data rate is below the bandwidth limit throughout the file.

Transitions

Transitions are effects that create a smooth transition between the ending of one clip and the start of another. Common transitions include fades, wipes and dissolves, all of which we'll look at in a moment. Many books have been written on when to use which transition and we have little to add from that viewpoint. Our focus is this.

From a compression perspective, transitions are "motion," which either spikes the data rate or forces the compressor to dynamically drop quality to reach the target data rate. So we did some analysis to identify "compression-friendly" transitions.

We used the Digital Video Producer from Asymetrix to create a transition from the meeting clip shown on the left of Fig. 11.13 to the talking head clip shown on the right. We used ten different transitions, outputting the entire sequences in raw form and compressing them with VidEdit using the Indeo Quick Compressor at a quality setting of 100 (Figs. 11.14 to 11.23).

We include both a picture of the transition and the frame profile, which shows whether the transition is neutral or if it adds motion that makes the compressor's life more difficult. We start with the transition that affects compression the least, and then illustrate progressively more compression-unfriendly transitions.

The transition starts on frame 15. You can judge compression-friendliness both on overall data rate and on whether the transition causes a "spike" in the general southerly direction of the data rate of this clip.

Figure 11.13 The two sides of the transition sequence, from left to right

Figure 11.14 The Wipe, the most compression-friendly transition, shows no blip at all on frame 15 and sports the lowest video data rate of the group

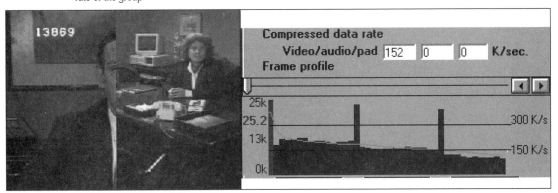

Figure 11.15 The Clock is also a solid performer from a compression perspective

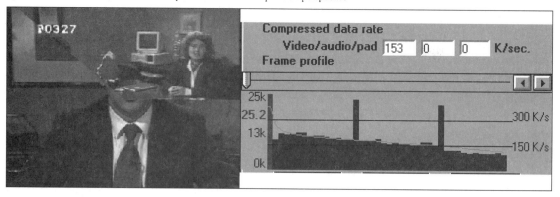

Figure 11.16 The Iris, also compression-friendly

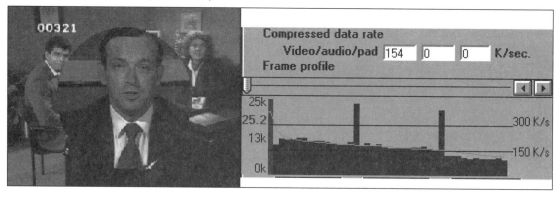

Figure 11.17 Blocks. We start to see a small jump in data rate and an interruption in the slope of the per second data rate line

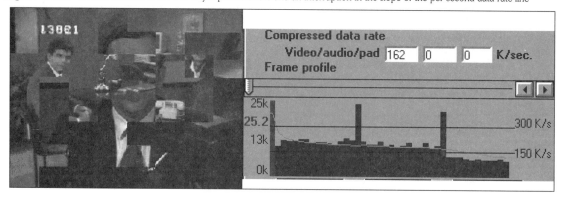

Figure 11.18 Blinds. Transition increases data rate slightly

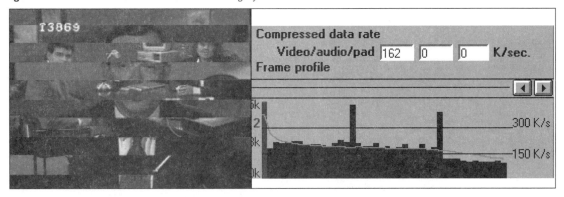

Figure 11.19 The Dissolve, a fairly common transition, changes every pixel in each transition frame. The codec sees this as motion and boosts the data rate, causing a big data spike during the transition

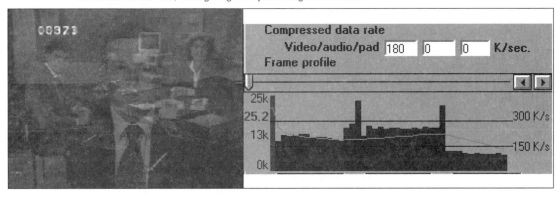

Figure 11.20 Bands, a favorite on the Love Boat, creates a large data spike and hikes overall data rate

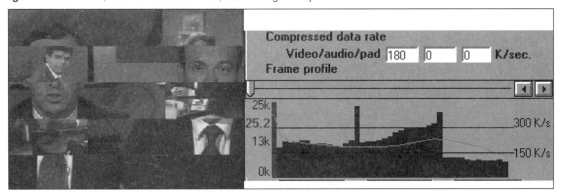

Figure 11.21 The Slide-In. This looks like the Wipe, but in the wipe both images are still and the line moves from left to right, covering one and revealing another. In this transition, the talking head *pushes* the meeting out of the frame so all pixels move every frame. The results are predictable.

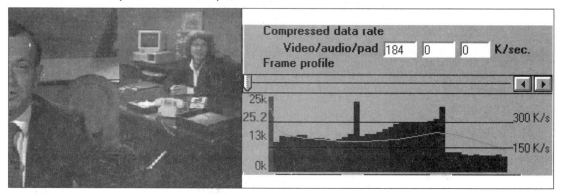

Figure 11.22 The Barn Door. The new video pushes the old out of the frame, creating a significant early spike and big increase in data rate.

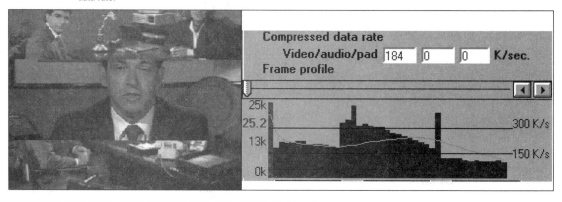

Figure 11.23 The Fizzle, appropriately named, was the least compression-friendly by a wide margin. Pretty funky looking, too!

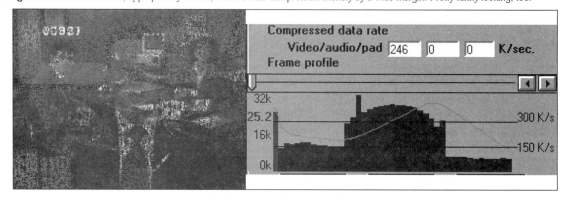

Summary

Overall, the friendliest transitions leave large portions of the frame static during most of the transition. They hide the old video and reveal the new, which limits interframe motion. The Wipe, Clock and Iris fit in this mold and are all relatively compression-neutral. Transitions that slide large sections of the frame create problems, as do dissolves and fizzles that affect the entire frame throughout the transition.

Most of the time, your higher-end codecs can compress these transitions down to the target data rate—that's not the problem. The price you pay is video quality, because that's what the codec trades off to meet the data rate target.

Fade-In/Fade-Out

How about fading in from black at the beginning of the sequence and fading out to black at the end? These sound close to the dissolve in operation; let's see if they have the same effect (Figs. 11.24 and 11.25).

Once again, a frame profile is worth a thousand words. The Fade In from black creates a significant data spike. The Fade Out to black is just as bad, but since it happens at the end of a sequence, it creates less of a problem.

If you're using the Indeo Quick Compressor, you may have to eliminate the word "fade" from your vocabulary. With all codecs, you should avoid fading to black at the end of one sequence and fading from black to start the next. The spike will put the codec in overload and video quality will suffer.

Hardware Requirements

For the most part, when working with video editors, you'll be dealing with raw digital video files, which are extremely large and bulky. This translates into huge RAM and disk drive requirements, but especially RAM.

Figure 11.24 Fading out to black. I don't know why I look so smug, from this angle the fade to black looks like a really bad idea

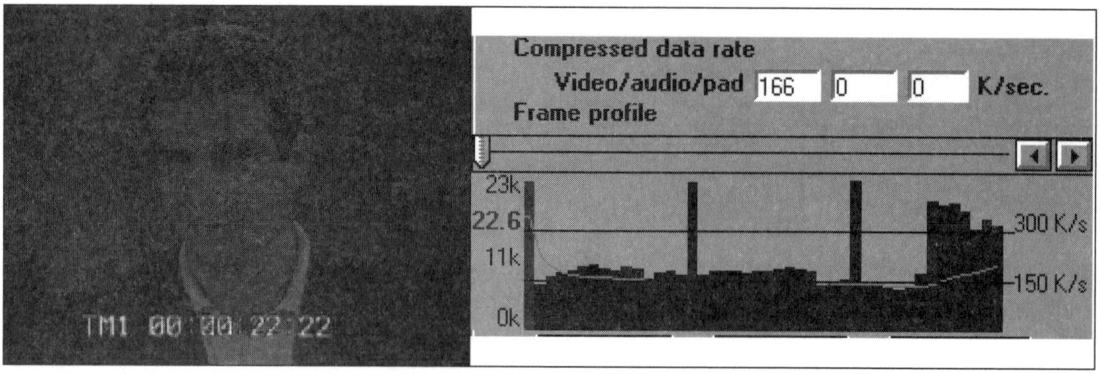

Figure 11.25 Fading in from black. Note the data spike that starts to moderate at the end of the fade

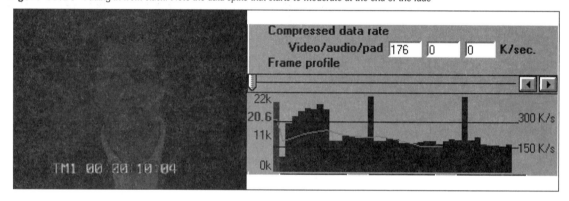

At eight Megabytes, all video editors page back and forth to disk so frequently you'll feel your computer shake. Literally (and you know I don't use that word lightly)! However, at 16 megabytes, they catch their stride, and performance seems positively zippy.

If you plan on doing significant video editing, take this recommendation to heart. Make sure you have at least 16 megabytes of RAM.

SUMMARY

1. VidEdit is the tool of choice for quick video nips and tucks. When clipping excess video frames, be sure to leave frames before the start of the clip for transitions in, and

frames at the end of the clip for transitions out. Remember to select "no recompression" before saving the file.

2. If you're dealing with relatively low-motion video files, the Doceo Filter can reduce the overall data rate and improve video quality.

3. When compressing with video editors, be sure to check their work against VidEdit, from both a display-rate and a data-rate perspective. Compression is a big black box, and few developers get it right the first time.

4. Transitions are neat, but some are easier to compress than others. Here's our summary:

   (a) Compression-friendly, use at will—Wipes, Clocks and Iris

   (b) Use sparingly, but only with Indeo 3.2 or Cinepak— Blocks, Blinds, Dissolves, Bands, Slide-Ins and Barn Door

   (c) Avoid at all costs—The Fizzle.

5. *Entry and Exit Transitions*—Fading in from black and out to black is almost as hard on compression as the Fizzle. Avoid these transitions if possible, especially when working with the Indeo 3.1 Quick Compressor.

6. If you plan on working with video editors, and only have eight megabytes of RAM, you have two choices. Spend $400 on tranquilizers you'll need to assuage your frustration level, or spend the dough on eight more megabytes of RAM. We strongly recommend the latter.

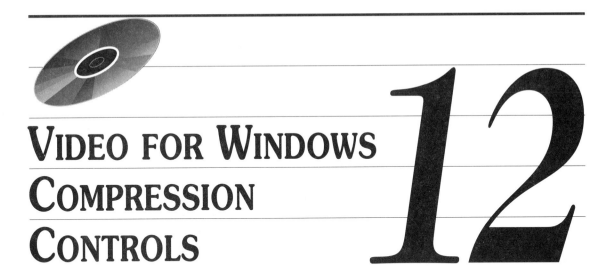

VIDEO FOR WINDOWS COMPRESSION CONTROLS 12

Now that our video is captured, pre-processed and ready for compression, it's time to select the codec and compression parameters. This chapter will outline selection criteria and alternatives. After reviewing the individual compression parameters, we'll work through four case studies, to illustrate in detail some critical Video for Windows functions.

Not considered here are frame rate and compression resolution, which are decisions most appropriately made during capture, not compression. These topics were addressed in detail in Chapter 10, Video Capture.

To make this chapter as self-contained as possible, we'll review some findings from previous chapters. We'll do our best to minimize redundancy.

Meet the Enemy

Figure 12.1 illustrates the *summary* version of VidEdit's Compression Options screen. Touch the "Details" button to open the complete set of controls illustrated in Fig. 12.2.

Figure 12.2 represents your sole link to the inner workings of the individual codecs. We'll review the various compression options from top to bottom, starting with Target.

TARGET

Target is a convenient mechanism to select several compression parameters at once. There are six settings: three for hard drives, two for CD-ROMs and one custom selection (Fig. 12.3).

Figure 12.1 VidEdit Compression Options screen—summary

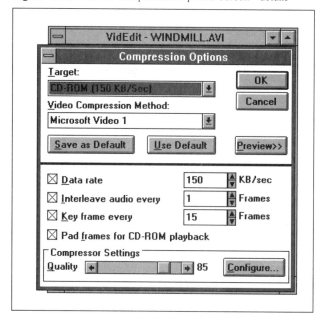

Figure 12.2 VidEdit Compression Options Screen—details

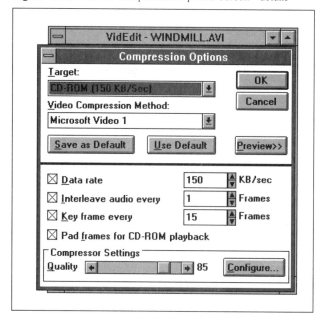

When you select the target, you set target data rate and also whether the files will be padded for CD-ROM playback.

For example, when you select CD-ROM (150 kB/sec) as the target, the data rate amount is set at 150 kB/second and the CD-ROM padding is enabled. If you select Hard Disk (150 kB/sec), the data rate is the same, but CD-ROM padding is disabled. We'll discuss CD-ROM padding in a moment.

All individual settings can be modified after you select the target. As we just learned, you expose the individual settings by clicking on details.

CODEC SELECTION

Preliminaries

No Recompression—We've mentioned the No Recompression option twice before: in Chapter 4, when we discussed Sound Synchronization, and Chapter 11, when looking at VidEdit. Not to beat a dead horse, but here it is again. Briefly, every time you load a file in VidEdit, the compression options used to compress the file are loaded into the Compression Options screen. When you save the file, VidEdit automatically begins recompressing the file to those parameters.

As we discussed earlier, you should compress each video as few times as possible, preferably only once, or twice if you

Figure 12.3 VidEdit Compression Options Screen—Target

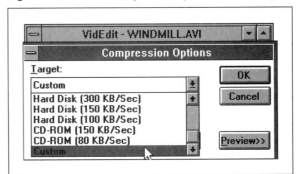

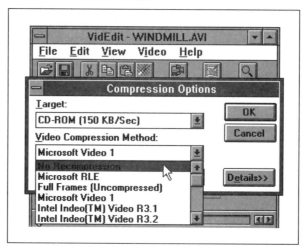

Figure 12.4 No Recompression Option in Codec Selection Screen

compressed during capture. Unless you intend to compress, select No Recompression before saving the file.

Full frames (uncompressed)—Also visible in Fig. 12.4 is the Full Frames (Uncompressed) option. This option converts the compressed file back to a raw file, expanding your tightly compacted 150 kB/second file back to 3.5 MB/second.

You probably won't use this option frequently, and the most important concept to remember is that Full Frames creates a completely different result than No Recompression. Unlike No Recompression, this option is always available, and it's often easy to get the two confused.

Selection by Footage

As we've seen, different types of videos impose different burdens on the codecs, and each codec has particular strengths and weaknesses that relate to video type. As a result, many publishers use different codecs for different videos in their titles. In the *Guide to Video Compression,* for example, we used Indeo for all 240x180 talking head sequences, Cinepak for motion sequences and RLE for screen sequences.

Chapter 6 completely analyzes the various codecs by video type, complete with pictures and sample videos. Here's a quick refresher course to help you select the best codec for each video type.

HIGH-MOTION SEQUENCES

Most codecs rely on interframe compression for the bulk of their compression. Since high-motion sequences have little of the interframe redundancies that fuel interframe compression, these sequences are particularly demanding. We evaluated the codecs based upon two primary characteristics, video quality and display rate. The latter characteristic is particularly important, because unless you display close to 15 fps, motion sequences appear choppy.

Cinepak beats Indeo 3.2 in quality, but just barely. However, Cinepak is still the most fleet-footed codec, outdistancing the pack by a wide margin. Video quality for both Video 1 and RLE quickly degrades in motion sequences.

Consider Indeo 3.2 for motion videos that are less than 320x240 in resolution. We've found that Cinepak files can become mottled at smaller resolutions, and reports from other developers match our own observations.

LOW-MOTION SEQUENCES

Low-motion sequences are relatively easy for the codecs. Here the focus shifts primarily to video quality, with display rate of secondary concern. At 320x240 resolution, the primary battle is between the Indeo 3.1 Quick Compressor and Indeo 3.2, which both typically producing higher video quality than the other codecs. We'll explore how to select and use these two options later in the chapter.

Consider Cinepak only if you value display rate over video quality. Consider Video 1 when your primary target platform is 8-bit display and palette considerations are important to your final program. To use Video 1 in 8-bit mode, you have to first convert your video to 8-bit mode. We'll describe how to do this in one of the case studies.

At smaller resolutions, Cinepak is a particularly poor choice, with low-motion videos appearing mottled. We've experienced mixed results with Video 1 at smaller resolutions; sometimes the videos look good, other times they don't. Once again, if 8-bit is your primary target platform, you might try Video 1 and fall back to Indeo if video quality isn't sufficient.

ANIMATION

Many publishers convert animated sequences into Video for Windows' AVI file format to interleave and synchronize sound with the animation. Synchronization is difficult or impossible with animated files in their native format.

Animated typically have low motion content, which opens the playing field for RLE and Video 1. These 8-bit codecs offer superior palette-handling capabilities and are a particular favorite for commercial publishers targeting 8-bit environments. If you find that these codecs can't handle the motion, try Indeo 3.2.

SCREEN CAPTURE

Video for Windows includes ScreenCap, a screen capture utility that allows you to capture still images or screen sequences. This is an extremely useful way to illustrate screen commands and command sequences, since you can display the exact command sequence with accompanying audio. Most screen sequences in the *Guide to Video Compression* were captured with ScreenCap. These files are all automatically stored in RLE format, which works quite well in this niche role.

DATA RATE

Factors in Data Rate Selection

All relevant codecs are lossy in nature, so as data rates decrease, so does video quality. This makes data rate selection a key decision, since it will ultimately affect the quality of your video as well as your storage space.

CD-ROM Publishers

Obviously, if your video will ship on CD-ROM, the decision becomes a bit more critical, because your data rate selection can also limit your target market. Today, most CD-ROMs being sold are double speed CD-ROM drives with around 300 kB/second capacities. However, single-speed 150 kB/second drives predominate the installed base.

A growing number of publishers are shipping "2x" or double-spin products that won't run on single-spin drives. However, our experience is that if you carefully capture and preprocess your video, 300 kB/second video doesn't look much better than 150 kB/second video, especially considering that you dramatically limit your target market.

Also worth noting is that as data rates increase, so does the processor overhead associated with retrieving the data. It makes sense—twice the data, twice the load on the CPU. Since CPU processing power is a zero-sum game, if data retrieval takes more CPU clock cycles, something else suffers. In digital video, usually that "something else" is decompression speed.

The drop in frame rate associated with higher-bandwidth files is illustrated in Table 12.1. Since the test bed was a rela-

Table 12.1 Tests performed on an 80486/66mHz clone, with VESA Local Bus and ATI Mach 64 Graphics Card with 2 MB of Video Memory. Note that Indeo 3.1 was tested, not Indeo 3.2.

	\–150KB	/S file–\	\–280 KB	/S file–\
	Processor Overhead	Display Rate	Processor Overhead	Display Rate
Cinepak High Motion File				
Hard Drive	19.2%	15 fps	37.2%	15 fps
CD-ROM	30.4%	15 fps	49.4%	14 fps
Indeo 3.1 High Motion File				
Hard Drive	19.2%	13 fps	37.2%	10 fps
CD-ROM	30.4%	12 fps	49.4%	8 fps

tively powerful computer, Cinepak's display rates didn't suffer dramatically. However, display rates for Indeo 3.1 dropped by close to 30%.

On slower computers, expect display rates for both codecs to drop by roughly 20–30%, the percentage of processing power stolen from the decompression and display task to retrieve the extra data. Obviously, the quality provided by the additional video bandwidth is fool's gold if it slows the display rates on your target computer. Keep this in mind when specifying the display rate for your CD-ROM–based product.

Speaking of fool's gold, let's compare the quality of 150 kB/second and 300 kB/second files. Obviously, the increase will directly relate to the amount of motion in the original file—the greater the motion, the greater the quality afforded by the additional bandwidth. For example, compressing a very low-motion talking head shot with Indeo 3.1 at target data rates of 150 kB/second and 250 kB/second yields files of virtually identical bandwidth, each around 134 kB/second. This means that the compressor reached maximum quality for that video at that bandwidth, and like skinny people I so very much envy, couldn't gain weight if it wanted to.

When you compress a high-motion sequence with Indeo at the same target data rates, the higher data rate settings yield a file approximately 92 kB/second larger (239 kB/second vs. 147). Ditto for Cinepak, which yields a file approximately 143 kB/second larger (284 kB vs. 141). Once again, the question becomes just how much quality the extra bandwidth buys you. We'll study that answer by comparing the 284 kB/second Cinepak file with its slimmer sibling (Figs. 12.5 and 12.6).

Sure, the higher bandwidth video looks better; how noticeable will it be at 15 frames per second? Is it worth losing the single-spin market? Will the increase in quality offset the 20–30% hit to display rate that the additional bandwidth may cause? Is it worth the CD-ROM real estate lost to the additional bandwidth?

Keep in mind that this video is extremely high-motion and therefore somewhat of a worst-case scenario. Videos with less motion will achieve even less quality improvement at rates over

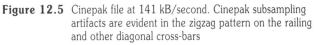

Figure 12.5 Cinepak file at 141 kB/second. Cinepak subsampling artifacts are evident in the zigzag pattern on the railing and other diagonal cross-bars

Figure 12.6 Same file at 284 kB/second. Note that zigzag artifacts on railing and diagonal cross-bars are gone. You can now almost recognize that this is a roller coaster scene. Unless, of course, you own a single-speed CD-ROM drive, in which case you can't run the video at all

150 kB/second—yet "a miss is as good as a mile," and if you're over 150 kB/second, you're out of the single-spin market.

If quality is your most important criterion, you might consider reducing video resolution. A 240x180 file video contains about 45% less screen area than 320x240, yet looks almost as large (Fig. 12.7). At the same 150 kB/second bandwidth, the smaller file will have 45% less data to compress, which should translate to 45% higher video quality.

Anyway, by now you've probably guessed that I lean towards the 150 kB/second limit. If you're faced with this decision, do the work—compress at the higher quality settings, measure the display rate hit and see if it's worth it.

When compressing for CD-ROM playback, remember that data spikes *anywhere within* the video file can interrupt smooth video playback. Thus, it's important to use VCS Play to perform a frame profile on all videos destined for CD-ROM playback. We covered this in detail in Chapter 5, Playback Platform Considerations.

Figure 12.7 Same video at 150 kB/s and 240x180 resolution. Note the absence of Cinepak artifacting found in the 320x240 150 kB/s video, and that the video is about the same quality as the 284 kB/s video compressed at 320x240 resolution. Of course, this video will play from a single-speed CD-ROM drive

Hard Drive Video

Obviously, if you're distributing your video on a hard drive the decision is not as black-and-white. Still, higher-bandwidth videos will slow display rate and consume hard drive space. The decision can become extremely sensitive in network and other communications situations, whether it's asynchronous dial-up modems or Lap Link transmissions between your compression station and the target playback computer.

On the other hand, if you're capturing for immediate playback, and capturing at around 300 kB/second, it probably doesn't pay to compress again unless you absolutely have to. Either way, you know the trade-offs and can make an informed decision.

Operation

The Compression Options screen contains two controls that are seemingly at odds, the Data Rate control (Fig 12.8) and the Quality control. We know that when you cut data rate with a lossy technology, you also cut video quality. So what happens, for example, when you set the data rate at 150 kB/second and Quality at 100, as in Fig 12.8, and the codec can't deliver the quality at the requested data rate?

Well, until Video for Windows 1.1, quality controlled. During compression, if the codec couldn't deliver the video at the requested data rate and/or quality setting, you got the error message shown in Fig. 12.9. Note that you don't get the option to reduce quality to achieve the requested data rate.

Of course, if you were publishing for CD-ROM distribution, the higher-bandwidth file did you absolutely no good. You had to restart the compression, this time at a lower quality level. How low? It was all trial and error, and all video- and codec-specific. So it was fairly easy to waste an entire afternoon compressing one or two files.

The public outcry over this situation had three key results. First, and most important, the codec developers prioritized data rate over quality. Today, every codec except for the Indeo

Figure 12.8 Data rate control. What happens when data rate is too low to achieve the target quality setting?

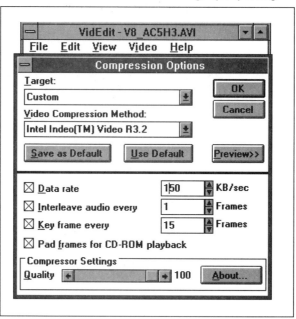

Quick Compressor, prioritizes data rate over quality. During compression, these codecs attempt to deliver the requested data rate, even if quality suffers. In fact, the Indeo 3.1 Super-Compressor, Indeo 3.2 and Cinepak all ignore the quality setting in most all instances.

Figure 12.9 Error message from Hades

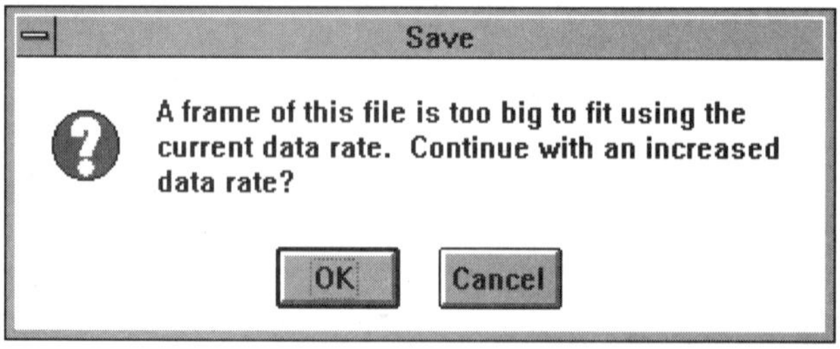

Second, both Indeo and Cinepak both improved their compression algorithms to the point that they could guarantee the requested data rate in most reasonable instances. Also a pretty cool result.

But the third response to the outcry over the disturbing error message is that Cinepak and Indeo stopped generating the error message. If the codec can't deliver the video file at the requested bandwidth, it just does the best it can and doesn't tell you that the file you just spent two hours compressing averages 154 kB/second. It's almost passive–aggressive, the codec's way of saying, "You're being unreasonable, and I'm not rising to the bait. Feed me some reasonable parameters and I'll do my job." Well, I've tried talking to my codecs, but that didn't work—still got the same result. However, as mentioned, with the recent technology advances, most of the time if you request a reasonable data rate, like 150 kB/second, you'll get it.

VFW Tip—When compressing high motion videos, all codecs tend to exceed the requested data rate by 2–3 kB/second. Since in fixed bandwidth applications "a miss is as good as a mile," it's best to request a target data rate about 3–5 kB/second below your real target. That usually ensures that you meet your target.

AUDIO

From our perspective, audio bandwidth is a "capture" parameter, or one best set during capture. See Chapter 10 to review factors involved in selecting audio bandwidth. You can also modify audio parameters using the control shown in Figure 12.10.

From a compression perspective, audio bandwidth is a component of total video bandwidth. Since data rate is another zero-sum game, the higher the audio data rate, the lower the video data rate.

We've duplicated Table 12.2, which illustrates the data rates at the various audio parameters. As you can see, at the higher-quality sound settings, the results are somewhat alarming, especially if you're working with CD-ROM products. Choosing the right combination relates primarily to the sound system on the

Figure 12.10 This VidEdit control
changes audio formats

target computer. Obviously, if you can't hear the difference, the additional quality isn't worth the bandwidth.

Whatever combination you chose, make sure you test on a computer with a very low-end sound board, which often yields some interesting results. For example, I've experienced a fuzzy sound called "white noise" when playing 16-bit audio on an 8-bit sound systems. Other developers have reported audio distortion when high-quality audio is played on lower-quality sound systems.

For these reasons, and to save bandwidth, most CD-ROM publishers seem to use either 22 or 11 kHz, 8-bit mono audio. All of the videos on the CD-ROM use 8-bit, 11 kHz mono audio.

Table 12.2 Audio data rates at specified audio parameters

Frequency	Sample Size	Mono Audio 8-bit	16-bit	Stereo Audio 8-bit	16-bit
11.025 kHz		11KB/S	22KB/S	22 KB/S	44 KB/S
22.05 kHz		22 KB/S	44 KB/S	44 KB/S	88KB/S
44.1 kHz		44 KB/S	88 KB/S	88 KB/S	176KB/S

Audio Compression

Video for Windows 1.1 introduced several forms of audio compression, including Microsoft ADPCM and IMA ADPCM. We haven't worked with either of these formats, but my biggest concern would be that the audio decompression overhead would slow your video display rate. You can test this by compressing the same video file twice, once with a standard WAV file and once with compressed audio. Decompress both and compare the display rates.

I'd be interested in hearing the results of any testing that you have performed. Now on to Audio Interleave.

AUDIO INTERLEAVE

Finally, a simple category.

The audio interleave control (Fig. 12.11) dictates how frequently audio chunks are divided for storage within individual

Figure 12.11 Audio interleave control

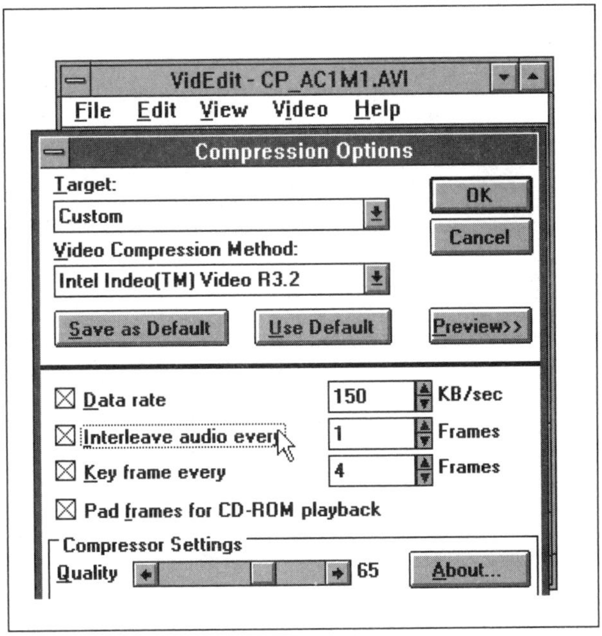

video frames. In a video compressed at 15 frames per second, for example, a setting of "1" will interleave 1/15 of a second of audio with every video frame. A setting of "15" will interleave one second of audio with every fifteenth frame.

A one-to-one interleave promotes a smooth file without data spikes, which is essential to smooth CD-ROM playback. You can use larger settings for files destined for other storage media so long as the interval isn't the same as the key frame setting. On the other hand, I've never heard any reason to use any value other than 1, so we use 1 all the time.

KEY FRAME SETTING

As you recall, key frames are the references for delta frames that aren't compressed during interframe compression. After much experimentation, we learned that the key frame interval affects everything from video quality to display rate. This section will discuss our findings, and describe how to choose the optimal key frame setting for your video and application.

Table 12.3 summarizes our recommendations regarding key frame setting.

High-Motion Sequences

As interframe motion increases, it becomes more difficult for frame differencing to keep up with the changes. By ignoring differencing, and focusing only on the frame being compressed, key frames let the compressed video stream "catch up" with

Table 12.3 Recommended keyframe settings

Video Type	Cinepak	Indeo	Video 1	RLE
Low Motion	15	15	15	1 per video
High Motion	7	4	15	1 per video
Animation	N/A	15 (max)	1 per video	1 per video

Figure 12.12 8.15 kB Delta frame in high-motion sequence compressed with Cinepak at key frame setting of 7 (default). Note the Cinepak subsampling artifacts on railings and cross-bars

Figure 12.13 The next frame is a 16.3 kB frame which clears up artifacting

 the analog video. This is illustrated in Figs. 12.12 and 12.13, which show a delta frame and a key frame in a high-motion sequence. Of course, the additional quality comes at a price, usually an increased data rate of about 10 kB/second.

As we saw in the Tour de Codec, Cinepak will create key frames when necessary to preserve video quality. Notwithstanding this innovative feature, however, you will achieve the best results at lower key frame settings.

Low-Motion Sequences

In low-motion sequences, the situation tends to reverse, and frequent key frames tend to detract from video quality. In low-motion videos, delta frames are extremely efficient because interframe changes are small (Fig. 12.14).

Key frames can "stress" the video, as shown in Fig. 12.15. Key frames also tend to shift the entire picture minutely, which can introduce a "flashing" effect into the video.

Unfortunately, Indeo, your codec of choice for low-motion videos, limits you to a maximum key frame interval of 15. If

Figure 12.14 8.15 kB Delta frame in a low-motion sequence compressed to around 100 kB/second with Video 1. While the image is a bit grainy, it still looks pretty good considering the extremely low data rate.

Figure 12.15 The next frame is a 14.2 kB/second key frame. Note the checkerboard effect created by this low-bandwidth frame. Subsequent delta frames pick up these artifacts and carry them through six or seven frames. In low-motion sequences, higher key frame intervals are preferred, because key frames don't add quality, they simply stress the codec.

you select a higher value, all Indeo versions will adjust the value back to 15 before compressing. When using Cinepak or Video1, however, higher key frame settings do produce a smoother image with fewer background shifts and flashes.

Degree of Interactivity

Low key frame intervals accelerate "seeking" to a random spot in the video file. This is because codecs can only seek to a key frame, not a delta frame. For example, if a video file had a key frame interval of 60, and the user wanted to seek to frame 118, the decompressor would seek to key frame 60, and then decompress frames 61–117 to build 118. While this is faster than actually playing the video, because the frames are not displayed, it can create a noticeable delay. For this reason, if users will frequently page through the video file, shorter key frames are recommended.

Special Situations

ANIMATIONS
As we've discussed, many producers convert animated sequences to AVI files to synchronize audio with the animation. We'll look at this process later in the chapter. For converted animations, the codecs of choice are Video 1, RLE, and Indeo 3.2, in that order. When using Microsoft RLE, the key frame setting is grayed out, and you automatically use one key frame per video (see Fig. 12.16).

RLE is offered by other vendors, including ATI Tech-nologies, and some let you modify the key frame setting. However, with animated files, you should always use one key frame per video, which means selecting a key frame setting equal to the number of frames in the video. For example, if a video had 312 frames, you would use a key frame setting of 312. Use this approach for Video 1 as well. Finally, when using Indeo 3.2,

Figure 12.16 Microsoft RLE automatically uses only one key frame for the entire animation

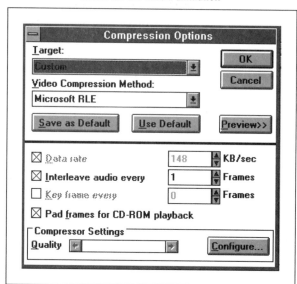

use a key frame setting of 15, the maximum allowed by Intel's codec.

SCREEN CAPTURES

We've discussed Microsoft's ScreenCap application, which enables the capture of both still and moving screen images and we'll detail how to use the application later in this chapter. As noted earlier and in Fig. 12.16, Microsoft's RLE, the codec used by this application, forces a key frame setting of one per video. If your RLE codec lets you modify the key frame setting, make it equal to the number of frames in the video file to ensure that you have only one key frame for the entire sequence.

PAD FOR CD-ROM PLAYBACK

Quick Answer

Most major CD-ROM publishers, including Microsoft, pad video files destined for playback on CD-ROM.

Discussion

CD-ROMs are divided into 2-kilobyte sectors. When CD-ROM drives read data, they must start at the beginning of a sector. This is called a "seek." Each seek can take from 200 to 500 milliseconds (as compared to 10–15 ms for fast hard drives). During seeking, the video flow to the computer stops, which endangers smooth video playback.

CD-ROM padding theory states that to allow the drive to "stream" and avoid seeking, all frames must begin and end on consecutive sector boundaries. If a new frame starts in the middle of the sector, the drive would have to stop, seek to the beginning of the sector, and start reading again.

When you enable padding, the compressor hands the compressed frame back to VidEdit, which "pads" the frame with garbage bits to bring the frame to a multiple of 2 KB.

Cinepak, Indeo 3.2 and Video 1 minimize CD-ROM padding during compression, producing CD-ROM padding figures of about 3–6 kB/s. Indeo 3.1 doesn't, and Indeo 3.1 files typically average about 15 kB/s of padding. RLE also doesn't minimize padding.

To Pad or Not to Pad

Video for Windows 1.1 includes improved caching and seeking strategies that appear to minimize CD-ROM padding. All CD-ROMs also provide some on-board buffering, and MSDEX, Microsoft's CD-ROM driver, also buffers. This buffering should smooth out the interruptions caused by multiple seeks. In addition, testing we performed failed to prove that CD-ROM padding helped performance, or that failing to pad hurt performance.

In their CompuServe Forum, Microsoft advised one user not to pad. However, a recent survey of CD-ROM publishers revealed that almost all still enable CD-ROM padding. These publishers included Microsoft, Video for Windows 1.1; Software ToolWorks, *Multimedia Encyclopedia 1.5*; Aris Entertainment, *MPC Wizard 2.0*; Compact Publishing, *Time Almanac 1993*;

Asymetrix, Compel; Eidlon, *Millennium Auction*. A survey of Microsoft's recent multimedia sampler disk, *Exploring Multimedia for Microsoft Windows,* revealed that most publishers represented on the disk also padded.

Following the pack, we padded on the *Guide to Video Compression* and the *Video Compression Sampler.*

QUALITY SETTING

Quick Answer

All codecs but the Indeo Quick Compressor—use 100.
Indeo Quick Compressor—see discussion.

Discussion

We've already discussed that before Video for Windows 1.1, the quality setting tended to control over the data rate, and that Video for Windows 1.1 now prioritizes data rate with the one exception listed earlier. For this reason, with the exception of the Indeo Quick Compressor, the best strategy is to set quality at 100 and select a data rate target two to three kB/second less than the real target to allow for slippage.

We discuss how to work with the Indeo Quick Compressor in the first case study. Note that while Indeo 3.2, the Indeo 3.1 Supercompressor, and Cinepak all ignore the quality setting, Video 1 does not, and will drop quality and data rate when quality is set below 100. You'll get the best results from Video 1 by setting data rate at your target bandwidth and quality at 100.

WINDOWS SETTINGS

Let's look at the setting that make a codec a codec under Windows. This involves another look at your system.ini file, which we first met in Chapter 9.

Figure 12.17 A never-before-published view of the drivers section of the author's system.ini file

```
[drivers]
VIDC.MRLE=MSRLE.drv    [Microsoft RLE]
VIDC.MSVC=msvidc.drv   [Video 1 driver]
VIDC.IV31=indeov.drv   [Indeo 3.1 driver]
VIDC.IV32=ir32.dll     [Indeo 3.2 driver]
VIDC.CVID=iccvid.drv   [Cinepak driver]
VIDC.KLIC=capcrnch.drv [Captain Crunch driver]
VIDC.RT21=indeo.drv    [Indeo 2.1 driver]
VIDC.YVU9=indeov.drv   [Indeo YUV-9 driver]
```

Figure 12.17 shows the line items that must be present in the "drivers" section to utilize the respective codecs. The listed drivers must also be loaded in your Windows\system directory to operate. Typically, either the Video for Windows installation program or the codec's installation program will modify your system.ini file and install the driver.

If your driver or system.ini file somehow get corrupted, the codec won't appear in VidEdit's Compression Options screen—you simply won't be able to find it. If the system can't locate or load the driver when loading or decompressing a file, you'll get the error message shown in Fig. 12.18.

If this happens, check your system.ini file for the appropriate line entry and make sure that the codec driver is present in

Figure 12.18 Error message that appears when trying to load a video compressed with a codec that is not properly installed

the Windows\system subdirectory. Sometimes your system.ini file will be inadvertently modified when installing or uninstalling capture boards, sound boards or associated drivers. It's pretty easy to fix if you know what to look for.

CASE STUDIES—PUTTING IT ALL TOGETHER

Now that the basics are behind us, let's pull it altogether and actually compress some files. We'll look at four basic scenarios, starting with a low-motion sequence to be compressed with the Indeo 3.1 Quick Compressor and Indeo 3.2.

LOW-MOTION—INDEO 3.1/3.2

As we've discussed Indeo 3.1 is made up of two codecs, a Quick Compressor and Super Compressor. Recently released Indeo 3.2 supersedes the 3.1 SuperCompressor, but eliminated the Quick Compressor because of space limitations in the build kit for Microsoft Chicago, or Windows 4.0. Fortunately, the Indeo installation is flexible enough to allow you to have both Indeo 3.1 and 3.2 installed simultaneously. If you installed Video for Windows with the enclosed CD-ROM, both codecs should be installed.

The 3.1 Quick Compressor ignores minor interframe changes. As a result, it compresses about 10 times faster than the Super Compressor, and when compressing low-motion video files, produces more compact files than the Super Compressor. By eliminating interframe noise, Quick-compressed video also looks better and displays faster than Super-compressed videos.

However, the side effect of minimizing interframe changes is inefficiency when interframe changes are significant. In other words, it can't produce low data rates with high-motion videos. For this reason, the Quick Compressor isn't a viable option for these videos—that's where the super compressor excels.

From an operational standpoint, the Quick Compressor is quality-driven, not data rate-driven. If it can't meet the target data rate at the requested quality setting, it just blithely presses on, delivering whatever bandwidth it delivers. Your only option is to compress at successively lower quality settings until you hit your target. However, since the Quick Compressor is extremely fast, this limitation is more of a hassle than a bar to use.

In addition, because quality rather than data rate controls the individual frame sizes, Indeo 3.1 can't optimize CD-ROM padding and averages about 12–15 kB/second. For this reason, we typically don't pad when compressing with this codec.

Notwithstanding these issues, we decided to use the Quick Compressor on the 240x180 sequences in the *Guide to Video Compression*. It performed admirably, producing high-quality video at data rates of around 100 kB/s, including audio. But let's do the work again, and see if you agree with our conclusions.

Accessing the Quick Compressor

The first step is to load the file into VidEdit, then select VidEdit's Compression Options screen. Accessing the Quick Compressor is relatively simple, though somewhat obscure. In VidEdit's Compression Options' screen, select Indeo 3.1 and toggle the data rate box so that it is empty rather than checked. This tells the Indeo 3.1 Codec to turn on the Quick Compressor. Figure 12.19 shows the proper configuration.

Now let's compress the same file using Indeo 3.2. Figure 12.20 shows the suggested parameters. The Quick Compressor crunches through about one frame every second. Indeo 3.2 takes about 10 seconds per frame. However, Indeo 3.2 delivers the requested data rate in most instances, where the Quick Compressor often forces you to recompress while attempting to control data rate by ratcheting down the quality setting.

Let's compare the two files (Figs. 12.21 and 12.22).

The pictures are close, but I give the still-image quality crown to the Quick Compressor in this instance. Where the Quick Compressor really shines, however, is in the almost total lack of interframe noise especially as compared to Indeo 3.2.

Figure 12.19 Suggested parameters for low-motion sequence compressed with the Indeo 3.1 Quick Compressor. Activate the Quick Compressor by unselecting the data rate box. Set audio interleave to one, key frames at 15. We don't pad for CD-ROM playback because the Quick Compressor doesn't optimize CD-ROM padding. Quality, which will control the overall data rate for this compression, is set at 100. If we don't meet our target data rate, we'll try again at a quality setting of 90.

Figure 12.20 Suggested compression parameters for low-motion sequence compressed with Indeo 3.2. Audio interleave is set at one, key frames at 15. Since Indeo 3.2 does a great job optimizing CD-ROM padding, usually about 3–4 kB/second, we elected to pad. The quality setting should have no effect on this compression. Nonetheless, we typically set it at 100. Just superstitious, I guess.

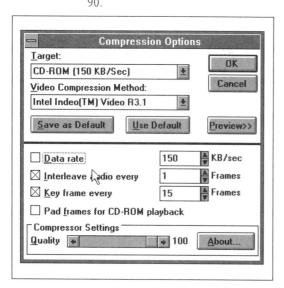

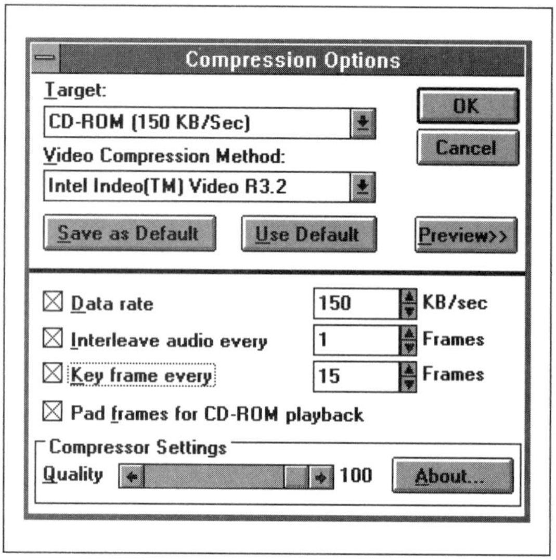

The reason why is illustrated in the frame profile shown in Fig. 12.23.

The frame profile reveals small spikes between the key frames on the left, which cause the noise we experienced with Indeo 3.2. The Quick Compressor ignores these differences and produces a much cleaner file. The data spike at the front of the Quick Compressor file tracks the fade-in transition in the beginning of the file. As we saw in Chapter 11, transition is a really bad idea when compressing with the Quick Compressor.

Figure 12.21 Indeo 3.2—video data rate of 120 KB/Second. The face looks a bit distorted, as evidenced by the flatness on the left cheek. Overall quality of the clothing and plants is very good. The back wall started out as a solid, smooth surface and was mottled during compression. When you play the video, you'll immediately notice the interframe noise manifested by continuous motion in the wall.

Figure 12.22 Indeo 3.1 Quick Compressor—video data rate of 139 kB/second. Overall quality is a bit better than Indeo 3.2. Where the Quick Compressor really shines is in interframe motion. When you play the video, you'll see the back wall shift only about once per second, which corresponds to the key frame setting. If Intel allowed unlimited key frame settings, this would be eliminated.

Figure 12.23 A comparative frame profile showing Indeo 3.2 on the left and the Indeo 3.1 Quick Compressor on the right

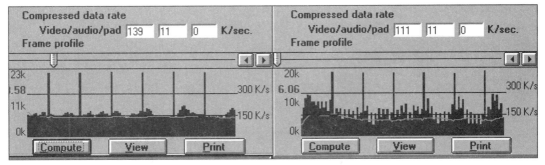

Conclusion

It's worth noting that we couldn't have achieved these results without first filtering the video. However, the Indeo Quick Compressor, used in conjunction with the filter, provides the absolute best video quality of all the codecs for low-motion sequences. Its fast compression time also makes it a valuable prototyping tool even for action sequences that it can't ultimately compress to CD-ROM data rates.

LOW MOTION— 8-BIT ENVIRONMENT— VIDEO 1

Suppose this video was destined for an 8-bit display, and after reading the Tour De Codec you decide to use Video 1 because it can hold a palette.

Now let's walk through the steps necessary to make this happen.

Converting to 8-Bit Format

Video 1 can compress in 8- and 16-bit modes, but you have to compress in 8-bit mode to achieve the benefits that we're seeking. The first step is converting the source video to 8-bit mode.

Figure 12.24 The wrong control for converting your videos from 24-bit to 8-bit, unless you're trying to create a gray-scale video

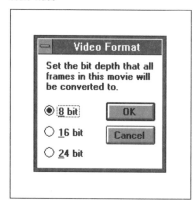

There's two ways to do this, the wrong way, which pretty much everybody tries six or seven times until they read the manual, and the right way. The wrong way is to use the Video Format control accessed from VidEdit's Video menu. This control shown in Fig. 12.24. It's a pretty natural mistake, given the subheading, but all you'll accomplish with this control is converting your video to gray-scale—a success of a sorts, since it is *8-bit* gray-scale, but a step backward nonetheless.

After staring at a gray-scale video seven or eight times, you'd start to feel like there's something you don't know, so you might check VidEdit's context-sensitive Help File from the Video Format dialog box. This is what you'd see:

```
When converting a 16- or 24-bit sequence to 8-bit
format, VidEdit remaps all colors to a gray-scale
palette. To retain accurate colors, you should not
use the Video Format dialog box to perform the con-
version. Instead, build an 8-bit palette (for more
information, see Creating Palettes), then apply the
palette to the entire sequence (for more informa-
tion, see Applying a Palette).
```

This makes sense, because the whole purpose of the exercise is to compress to standard palette, and we haven't told VidEdit which palette to use. In essence, the way Video for Windows converts from 24-bit to 8-bit video is to create an 8-bit palette and then paste the palette to the video.

Creating a Palette—Video

Creating a palette from the video itself is fairly simple. Use the Create Palette control found in Fig. 12.25 to create a palette for the individual frame, the entire video or any number of contiguous frames. The maximum number of colors that a palette can contain is 256, which includes 20 colors reserved by Windows. When you select 256 colors in your palette, you really get only 236, plus the 20 colors reserved by Windows. When you select a number below 256, VidEdit creates a palette containing the 20 Windows colors and the number you selected, up to a total of 256.

If the video is essentially one sequence without significant scene changes, it's fastest just to compute the palette from the current frame. It there are many different scenes, you should compute the palette from all frames. This can take a few

Figure 12.25 Control for creating a palette used to convert 16 and 24-bit videos to 8-bit

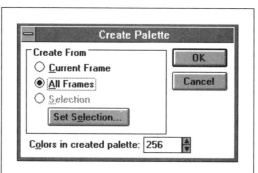

moments when your video is on a hard drive, and seemingly hours when the video is on a CD-ROM. Once the process starts, there's no stopping it except by shutting down your computer or performing the three-finger salute.

You can create an unlimited number of palettes for any video sequence, but during compression VidEdit will limit the total number of colors to 256. In operation, it's usually easiest to let VidEdit select the optimum palette for the entire video, then paste the palette to the entire video.

Pasting a Palette

The Paste Palette control shown in Fig. 12.26 lets you paste the palette into one frame, all frames or a selected number of continuous frames. Frames that the palette isn't pasted into will become gray-scale frames, although the color information can be regained by computing a palette as shown earlier.

Always select "Remap the video to the best palette colors," which arranges the new palette to match the colors in the old palette as closely as possible. Otherwise, the new palette is blindly mapped over the old palette, which typically causes color distortion.

Figure 12.26 Pasting a palette into a video file

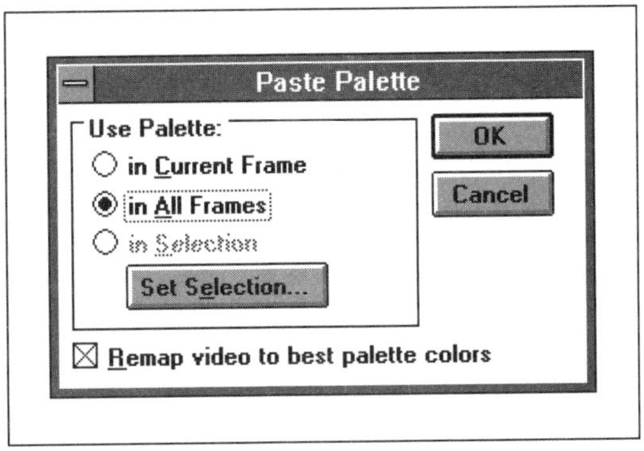

Once you've pasted the palette, the video is in 8-bit mode. When you compress this 8-bit file with Video 1, the final video will always decompress back into the palette used to create it.

Since you'll probably use Video 1 only for low-motion sequences, use a fairly high key frame setting unless the video will be used interactively. This will optimize quality over the length of the video and eliminates background "pulsing" often caused by key frames. Start with one key frame for the entire video. If you notice blurriness in the motion segments, drop the interval to around 60 and try again. The default key frame interval of 15 should only be used as a last resort.

The best data rate/quality strategy for Video 1 is to select about 5 KB/second *below* the actual desired data rate and use a quality setting of 100. This will lengthen your compression time but provide the highest possible quality of video for the selected bandwidth. Using the default setting of 75 will definitely lower the data rate and detract from overall quality.

Creating a Palette—Presentation

Typically, when developing for an 8-bit environment, video is only one of the graphic objects that needs consideration. Animations, bitmapped images and even buttons and icons must share the same palette, or your palette will flash when you display objects with nonconforming palettes.

Table 12.4 VidEdit's input formats

Source	File Extension
Truevision	TGA
Apple Macintosh	PIC
Compuserve	GIF
PC Paintbrush	PCX
Microsoft (Windows/RLE)	DIB
Microsoft	DIB Sequence
Autodesk	FLC/FLI

One little-known use of VidEdit is to compute an optimal palette for a group of disparate graphic types. Once you load a video, you can use the File/Insert command to load the formats listed in Table 12.4 into the video file.

As shown in Figure 12.27, VidEdit will even shrink or stretch the image to fit into the video, making it easier to add a disparate group of images into the palette computation. After you load all of the file types and different colors, use the

Figure 12.27 Using VidEdit to create a common palette for all graphics objects in your presentation

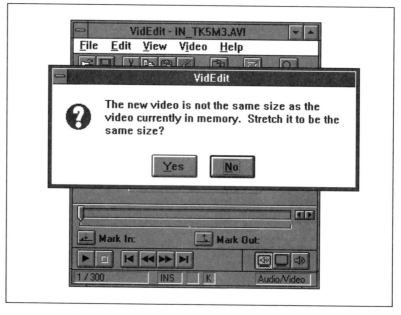

Figure 12.28 Pasting a palette into PalEdit for storage and later use

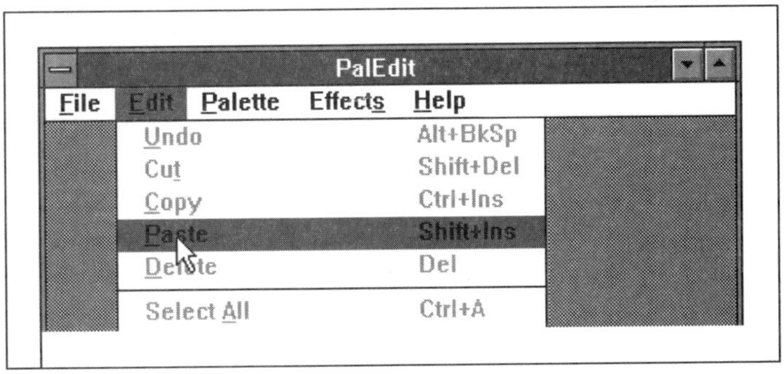

Create Palette command to create a palette for the image group. Then paste the palette to all video frames and look at each graphic image to make sure that the palette is acceptable.

VidEdit doesn't allow you to save a palette for later use. If you have PalEdit, load the program and paste the palette as shown in Fig. 12.28. Once you paste the palette, you can save it and apply it to other video and bit-mapped images in the future.

Note that VidEdit can't apply a palette stored on disk to a video file. VidEdit must either compute the palette, as shown earlier, or access a palette copied into memory by another program. PalEdit does a fine job at this task as well. Simply open the palette file, Select All (under Edit) and Copy the entire palette into memory. Once it is in memory, you can paste the palette into a video file with VidEdit or into a bit-mapped image with BitEdit or other bit-mapped editor.

If you don't have PalEdit, save any frame from the original video as an AVI file. Next time you need the palette, load the AVI file and compute the palette as discussed earlier. The computed palette should be identical to the previously computed palette.

ANIMATION CONVERSION— VIDEO 1

As we discussed in the Tour de Codec, developers often convert animated files to video to take advantage of the synchronization features of the AVI file format. Converting an animation into an AVI file is a simple process. As shown in Fig. 12.29, VidEdit can directly load animated FLI or FLC files from Autodesk Animator and Animator Pro.

Figure 12.29 Using VidEdit to convert an animated FLI file to AVI file format

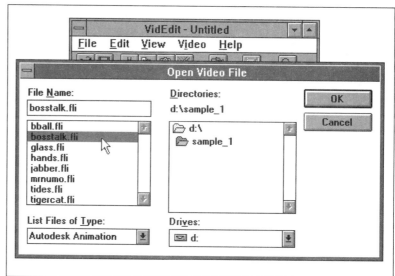

Once you load the animated file, VidEdit converts it to an AVI file. You can perform all of the normal Video for Windows editing functions and compress the file with any codec.

Note that VidEdit's conversion capabilities are strictly a one-way street. You can't take an AVI file and convert it back to FLI or FLC format.

If your computer has less than 16 megabytes of RAM, and your animation is close to full-screen in size, you could run into memory problems. These will manifest as either very slow compression that ultimately grinds to an ugly halt, or frequent "out of memory" errors.

You can sometimes cure the memory problem by displaying the animation at one-half scale. Use the Zoom control either through the menu controls (View/Zoom) or by clicking on the hourglass icon with your left mouse button.

If that doesn't work, extract the file as a DIB sequence (File/Extract) and then reload it into VidEdit. You should be able to compress normally. The memory problems seem primarily related to the animation display function and seem to disappear when working with the DIB sequence, even though the latter format is much more bulky.

Figure 12.30 Video 1 animation on a key frame

Figure 12.31 Same animation on a delta frame

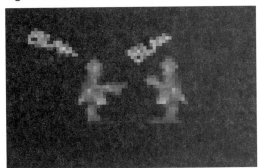 

When we converted and compressed the animated sequences on the CD-ROM, we created minor artifacts that you may or may not have noticed. However, we felt that synchronized AVI files with these artifacts looked better than perfect animations without synchronized audio. Also, AVI files are easier to work with, we didn't have to install the AutoDesk player. Maybe you won't reach this same conclusion, but here are the considerations.

As we mentioned back in Chapter 6, animations are "low motion" videos and key frame settings should be as high as possible—up to one for the entire sequence. Animations in particular show the negative effect of key frames, as shown in Figs. 12.30 and 12.31.

If the animation has a major transition, set your key frame on the transition frame. For example, if the transition is on Frame 52, use a key frame interval of 52. This will maximize the quality of the video.

For overall manageability, try the 8-bit codecs, Video 1 and RLE, first, in that order. Use Indeo 3.2 as a last resort, since you lose all your nice palette-handling capabilities. You may run into memory problems with RLE and Indeo 3.2 with videos larger than 320x240. Compress at your target data rate and a quality setting of 100 for these three codecs.

Finally, also as mentioned in Chapter VI, be mindful of bandwidth limits when working with animations. The first animations developed for the *Video Compression Guide and Toolkit* were elaborate, creative, swirly things with shifting backgrounds, moving lighting and 3-D motion for all objects. They were also 11 megabytes in size for about 20 seconds. You don't have to be Einstein to quickly see that that particular animation wouldn't play unless you had a quad-spin CD-ROM—hardly our target playback system. It took some ruffled feathers and scaled-back expectations to meet our bandwidth goal, and conversion to AVI format only makes the matter worse.

At a recent conference, I joked that animators creating for CD-ROM distribution should work in a room with a ceiling about 5 feet high. That way, every time they stand up they would bump their heads and hopefully remember that CD-ROMs are a fixed pipe that must be designed for and not ignored.

OK—they didn't think it was funny either. But neither is paying thousands of dollars for animations you can't use, or having to rein back the creative juices of your animation jockeys late in the game.

ScreenCap/ VidEdit

One of the coolest features in Video for Windows 1.1 is ScreenCap, the screen capture program for still images and video. Many screen shots in this book were captured with ScreenCap, and all of the example videos contained in our Guide to Video Compression were also captured with ScreenCap. Figure 12.32 and Fig12_32.avi are also examples.

ScreenCap captures all screen shots as 8-bit AVI files in RLE format. You can add audio to create training videos, or extract single frames through VidEdit. Before loading the program, configure your system in 8-bit mode, which improves the program speed and responsiveness. ScreenCap's 8-bit operation makes it sub-optimal for many still image screen captures, especially if video or 24-bit graphics is involved. You can, of course, copy these in 24-bit color format simply by pressing ALT and selecting Print Screen, which copies the current screen to the clipboard.

Once your computer is in 8-bit mode, you're ready to start. Double click on the ScreenCap icon to load the program. The

Figure 12.32 Frame 1 of a ScreenCap movie captured on disk

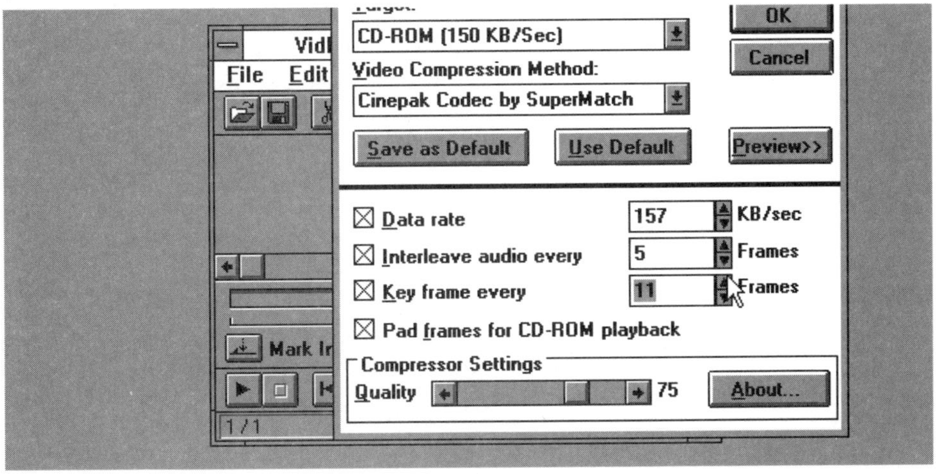

ScreenCap icon should appear in the bottom left hand corner of your screen. This means the program is resident and ready to capture.

By touching the icon, you access the following options.

SET CAPTURE FILE

Names the file to contain the captured footage. Note that ScreenCap will overwrite the file during consecutive captures unless you change the name each time or elect to automatically update the file name. If you elect this option, use four or five letters in the name followed by 001 (e.g., CAP001.avi). ScreenCap will automatically update the numerical portion of the file name each time you capture, preventing you from overwriting a captured file.

SET CAPTURE WINDOW

When you select this option, ScreenCap places a red and white box on the screen that you can resize with your mouse or set by selecting the top left pixel and width and height parameters.

Figure 12.33 ScreenCap preferences dialog box

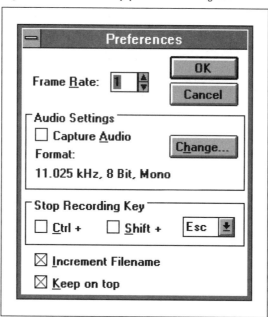

When capturing to extract a still image, either method works well, although you might want to standardize on one image size. When working with videos, standardization is particularly important, because you'll be creating playback windows for all videos and it's a real pain if they're differently sized.

With some video cards, if the video display window programmed into your presentation doesn't exactly match the video resolution, the display gets extremely distorted and looks like a monitor that has lost synchronization. This places a real premium on writing down the parameters of the capture window size and keeping a log of your efforts.

PREFERENCES (FIGURE 12.33)

Frame Rate   When capturing to extract a still image, use a setting of 1. When capturing moving video, use a setting of 5–10. We used 5 fps in the *Video Compression Guide and Toolkit*.

Capture Audio	We never captured audio with this program; we captured video and added the audio separately. Unless you're capturing some type of synchronized screen program, I can't think of any reason to capture audio and video at the same time. Narrating your own screen movie in real time will probably not be very successful—you're better off scripting it carefully, capturing the audio separately, and then synchronizing the two later.
Stop Recording Key	This is the key that stops the recording. Use the "escape" key unless it conflicts with the program that you're capturing from.
Increment Filename	Probably should always use this option just to guarantee that you won't overwrite your files.
Keep on Top	CapScreen disappears once you start your capture, so there's probably no reason not to keep it on top all the time.
Capture	Starts the capture which will continue until you hit the Stop Recording Key, or Escape if that's what you've configured. When you press Capture, the ScreenCap icon will disappear, and the cursor will convert into an hourglass, indicating that the computer is preparing for capture. When the cursor returns to normal, you're capturing.
Still Image Capture	When capturing to extract a still image, the capture frame rate should be set at one frame per second. This gives you plenty of time to "pose" the screen shot, moving your screen elements into the desired position. Make sure your cursor is located where you want it to be. Wait one or two seconds after you've finished posing, say "Cheese," and punch the Escape key to end the capture.
Video Capture	When capturing a series of screen movements, it's usually best to choreograph the mouse movements, or at least create a point-by-point outline. It's amazing how

difficult it is to smoothly move through thirty or forty seconds of mouse commands—almost like stage fright.

Remember that your viewer is looking at a menu or spreadsheet, not the Braves playing the Giants or Melrose Place. Fifteen seconds without motion seems like forever, even when accompanied by audio, so get in the habit of keeping the cursor moving as much as possible. For example, when talking about a button or control, don't point to it, circle it with your mouse. When describing a drag and drop operation, do it repeatedly, not just once. It's amazing how much this helps keep the viewer's attention on the video.

Most of the time, when you press a menu command that launches a freestanding dialog box, it will appear on screen straddling or totally outside the capture window. When this occurs, drag the box back into the capture window and keep capturing. During edit, you can cut out all frames showing the box outside the window and make your viewers think your software is perfectly behaved.

Copy Frame

Use this option when capturing still images for pasting directly into a program. Doesn't automatically create a file that you can save; you have to use another program to actually store the image.

Edit Captured Video

A convenient way to load the video into VidEdit, for extracting the still-frame image or editing the video file.

Editing the Video

All ScreenCap files are stored in RLE format at a key frame setting of one per the entire sequence. Use the File/Insert command to insert a WAV file or files containing the narration. You can do this at the beginning of the video file or from any frame in the file.

Typically you'll have to edit both the audio and the video to polish the final result. You can edit the video fairly simply with VidEdit's Cut and Paste commands (Fig. 12.34). To lengthen the video to accommodate additional audio, "cut" a frame and then "paste" it as many times as necessary. If you use five frames per second, like we did, five "pastes" equals one additional second.

When editing video frames, make sure that you're editing only video frames. Do this by selecting the "Video Only" track using "Edit/Track," or push the "Video Only" button on VidEdit's lower right-hand corner (see cursor location in Fig. 12.34).

On the audio side, you'll get the best results by recording in small chunks or sound bites, usually one or two sentences long. That way you can insert the audio where necessary without having to cut and paste the audio with VidEdit.

Typically, when you paste audio into a video file and then try to play the audio, audio playback will be choppy until you save the file. This is normal. Once you save the file, audio

Figure 12.34 Editing captured video with VidEdit

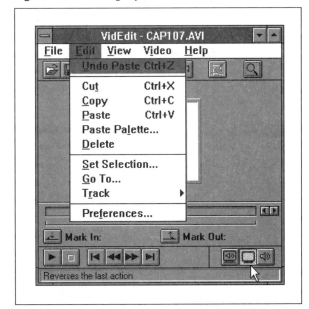

Figure 12.35 Extracting a frame from VidEdit

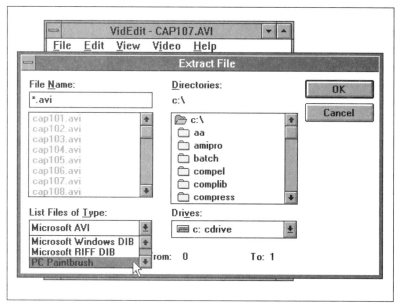

playback will normalize. Before saving the file, select "No Recompression" in the Compression Options Dialog Box to avoid recompressing.

Extracting a Bit-Mapped Image from VidEdit

When using ScreenCap to extract a bit-mapped image, the actual extraction takes place in VidEdit. Use the Edit Captured Video command to load the file into VidEdit. Scroll to the target frame, and select File/Extract (Fig. 12.35). The frame extracted will be the frame on screen when you press OK.

Supported export files are as shown in Table 12.5.

When you export a ScreenCap file, it's already in 8-bit mode so the color information follows the frame. However, what happens when you extract a frame from an Indeo or Cinepak file?

Table 12.5 VidEdit's export formats

Source	File Extension
Truevision	TGA
Apple Macintosh	PIC
Microsoft	AVI
PC Paintbrush	PCX
Microsoft (Windows/RLE)	DIB
Microsoft	DIB Sequence

Well, let's see. We saw before that if you use the Video format dialog box in Fig. 12.24, you convert the file to gray-scale, essentially because VidEdit doesn't know which palette to use. The same logic applies here. If you extract a frame from a 24-bit video file, you end up with a gray-scale frame.

To extract an 8-bit frame, you have to create a palette using the Create Palette dialog box shown in Fig. 12.25. Create the palette solely from the target frame and apply it to the target frame. Then extract away, and you've got an 8-bit bitmapped image. Once again, you may get better results using the ALT-Print screen combination to capture in 24-bit format.

SUMMARY

1. Compression Parameter Quick Reference Guide

   (a) Codec:
      i. Low motion—Indeo Quick Compressor unless 8-bit environment, then consider Video 1.
      ii. High motion—try Cinepak first, but keep Indeo 3.2 in mind.
      iii. Animation—Video 1, RLE, Indeo 3.2

   (b) Data Rate:
      i. CD-ROM-based video—150 kB/second
      ii. Drive-based:
         Low motion—150 kB/second
         High motion—improvements stop around 250 kB/second

When compressing for fixed bandwidth mediums, always request 3–5 kB/second less than your target bandwidth to allow for potential spillover, especially for high-motion videos.

(c) Audio Interleave—1

(d) Key Frame Setting:
   i.   Low motion—at least 60 unless interactive. Indeo limit is 15.
   ii. High motion—Cinepak (7), Indeo (4)
   iii. Animation—one per video

(e) CD-ROM Padding:
   i.   Indeo 3.2, Cinepak, Video 1—yes
   ii. Indeo 3.1 Quick Compressor—no

(f) Quality Setting—100

2. Use Indeo Quick Compressor for cleanest background. Access the Quick Compressor by unchecking the data rate box with Indeo 3.1 selected. Quick compressor is quality-driven, so you reduce the data rate by lowering the quality.

3. Steps in an 8-bit Video 1 compression:

   (a) Convert to 8-bit by creating and then pasting a palette.

   (b) Compress with Video 1.

4. VidEdit is great for computing a combined palette for a collection of videos, animations and bit-mapped graphics. Use PalEdit to store the computed palette.

5. ScreenCap is a wonderful application for building software training films and capturing still images. Keep it in mind for all your titles.

VIDEO FOR WINDOWS
DEVELOPMENT TIPS

13

VIDEO FOR WINDOWS DEVELOPMENT TIPS

In many ways, working with Video for Windows is like living in a haunted house. You can't predict when strange things will happen, but the longer you live there, the more likely some seemingly innocent step will lead to totally bizarre results.

Imagine, if you will, the 8-bit environment. A video-friendly environment with low decompression bandwidths accelerating playback for even the slowest codec. But then, you inadvertently apply the wrong palette and

SHRIEK!!!!!!

the video that played at 30 frames per second just a moment ago is now crawling at 10 fps. You can almost hear it gasping for air. Your computer performs like an old AT, and your fellow programmers start to edge away slowly with averted eyes. You've entered . . . the Palette Zone.

And then there's my favorite, the embedded vs. pop-up window. The day we sent VCS Play off to get reviewed by a major weekly magazine, we made the mistake of trying one more test and discovered that window size and type can affect playback rate by as much as fifty percent. Can you say panic?

I mean, what do you say when a reviewer points out that the same video plays at 10 frames per second in the right-hand window, and 15 in the left? It's a feature? Got a few gray hairs on that story, but luckily averted a catastrophe with some quick development work and, as always, a lot of luck.

So step into my parlor. We've got some fascinating stories to tell. Do stay a while.

THE PALETTE ZONE

The Problem

We originally addressed these materials in The Tour de Codec, but we've covered a lot of ground since then and a review will be helpful. Most Windows-based video graphics cards display in 8-bit, 16-bit or 24-bit color depth, with 8-bit systems predominating. As we've seen, most codecs work faster in 8-bit mode—especially native 24-bit codecs such as Indeo and Cinepak. This is because there's less data to display to the Windows buffer and less to pump from main memory to the video card.

Eight-bit displays are limited to 256 colors. This collection of 256 colors is called the palette, since all screen elements must be painted with colors contained in the palette. The 256 color combination is not fixed—palettes can and do frequently change. But at any one point, only 256 colors can be used to describe all the objects on the screen.

When displaying in 8-bit mode, all codecs are limited to 256 colors. For 8-bit codecs (Video 1, RLE, MPEG) this isn't a problem—their video is already described in 256 or fewer colors. This means that 8-bit codecs look as good in 8-bit mode as they do in the higher color depths. 24-bit codecs (Cinepak, Indeo) use more than 16 million colors to describe their true color video. To drop from 16 million to 256 colors, 24-bit codecs dither or draw geometric pixel patterns of various sizes to simulate colors not contained in the palette.

Palette Flashing

As we just noted, all 8-bit displays use a set palette. In general, when a new graphic object displays, the display palette changes to the palette of the new object. If that palette is different, the screen "flashes," momentarily blanking out and realizing the colors of the new palette. This detracts from the smoothness of the overall presentation.

Eight-bit codecs can "hold a palette," which allows developers to select one palette for a screen or presentation, and compress all videos and other graphics to that palette. Cinepak and Indeo can't hold a palette and always decompress to their own unique palettes unless otherwise commanded in Windows.

Last chapter we walked through the process of computing a group palette, converting a raw video to that palette and compressing with Video 1. However, if your video has even a moderate amount of motion, Video 1 will not perform well. Which gets us back to Indeo 3.2 or Cinepak.

In essence, there are two ways to avoid palette flashing. You can force these two 24-bit codecs to decompress to a fixed palette, or apply their palettes to all other screen objects. Let's explore the implications of the two approaches.

Decompressing to a Fixed Palette

Most presentations involve a varied group of graphic objects including icons, bit-mapped graphics, animations and, of course, video. The goal in palette design is to maximize the appearance of all the graphic objects, which can be tough given that you have to use the same 256 colors for all objects. As we discussed last chapter, VidEdit provides a convenient mechanism to build an optimal palette for these graphic objects.

Forcing the video to decompress to this palette is also fairly simple. If you're programming with MCI controls, you can use the setvideo_palhandle MCI command, documented on page F-10 of the Video for Windows User's Guide. Otherwise, an increasing number of authoring programs also comprehend this command through their controls.

This would seem to be the best approach—you can comprehend the video in computing the optimal palette, and use a palette that's well suited for all screen objects. However . . . you can almost hear the music . . . do de do do, do de do do . . . what happens when you force these codecs to decompress to a palette other than their own?

Consider this before you guess. Both Intel and SuperMac designed their palette to accomplish two goals, the first appearance and the second decompression speed. So what's the first thing that goes when you won't let them play with their own ball?

Yup. Decompression speed. Figure 13.1 shows what happens when you force Indeo and Cinepak files to decompress

Figure 13.1 Display rates in and out of native palettes

Test Description	Native Palette	Other Palette
Indeo 3.2—30fps 320x240 video—playback on 486/66 Gateway with ATI Mach 32 video card in 8-bit mode	30 fps	15 fps
Indeo 3.1—15 fps 320x240 Indeo 3.1 video—playback on 486/66 clone from double-speed CD-ROM drive with ATI Mach 32 video card	15 fps	10 fps
Cinepak—30 fps 320x240 on Gateway	30 fps	11 fps
Cinepak—30 fps 320x240 on clone	30 fps	10 fps

and display in a palette other than their own. I'm sure you'll find the results . . . quite strange.

You can easily perform these tests your self. Load act_in32.avi on the left-hand side of VCS Play. Play the file and record the display rate. Then load act_cp.avi on the right-hand side. The palette should flash as it changes to Cinepak's palette. Play the Indeo file again and the display rate should drop substantially. To try Cinepak, exit VCS and Windows, and start over, reversing the order of loading the file. Cinepak will slow down as well. I can't guarantee that this test will work on every video card, since the videos will often switch back to their palette before playing. In fact, it works on one of our ATI Mach 32 cards and not on the other. I *can* guarantee, however, that if you force the codecs to decompress to another palette, you will experience this drop in frame rate.

While performing these tests, note how awkward both codecs look in each other's palette. It was just this problem that Microsoft and the codec vendors sought to address with the Active Palette feature on Video for Windows 1.1c.

Active Palette

In Video for Windows 1.1, when codecs decompress to a fixed palette, they blindly assumed that palette, which created the bizarre results you just witnessed. For example, if color 14 in the assumed palette was navy blue, and color 14 in the codec's palette was red, your local boy scout troop looked like a communist youth organization.

Active Palette Technology, introduced with Video for Windows release 1.1c, lets codecs intelligently reorder their palette to match the system palette. In our example above, if color 124 in the codec's palette was navy blue, it could use that color in the place of system color 14. This makes all videos displayed in 8-bit mode more lifelike.

What we don't yet know is whether this will also allow the codecs to decompress and display faster than under the current scheme. If not, developers committed to Indeo and Cinepak have a real tough decision. They can force the optimal palette on the video, and slow display rate by up to 66%, or use Indeo or Cinepak's palette as the native palette for the entire presentation. This maintains the video display rate, but may detract from the overall appearance of the presentation.

Indeo and Cinepak Palettes

Indeo and Cinepak's palettes are saved in the chap_13 subdirectory under Indeo.pal and Cinepak.pal. You can also derive them yourself by loading a Cinepak or Indeo file in VCS Play, playing the video, and then press ALT and Print Screen. This copies the screen, along with the palette into the Windows clipboard where it you can paste it directly into PalEdit.

You can use BitEdit or other image editing programs to apply the palette to other bit-mapped images in your presentation. Animation files must be handled in the program that created them, such as Animator Pro.

Overall, using the native palette of the primary codec seems to be the simplest way to avoid palette flashing and maintain video display rates. It takes much of the art out of creating a presentation, but keeps you firmly outside of the dreaded Palette Zone.

MIND YOUR WINDOWS

A story.

Do you get nightmares when you ship a product to a reviewer or client? Do you finally grasp just how they'll test your products the second after you hand it to the FedEx person? Do you wake up in a cold sweat cursing yourself for not doing the one test so obvious that anyone else developing the product would have performed it before mastering 1,500 CD-ROMs?

Me, neither. But there was this one time. . . .

It was 7:00 PM. I had just FedExed a review copy of the Video Compression Sampler (VCS) to a prominent computer weekly magazine. As you know, one key feature of VCS is calculating the display rate of a video file. I was writing the reviewer's guide, to be sent via CompuServe that evening, and started writing the Competitive Analysis. VidTest, Microsoft's Video for Windows utility, performs the same function (although much less elegantly, of course), and it struck me that the reviewer would likely play a file with both programs to see if they yielded the same results.

So I did the same, and guess what? They didn't. VidTest was faster. And the reason why relates to two Video for Windows quirks that anyone using AVI files should know about.

VCS's dual display panel was designed for comparing files compressed with different technologies. We had informally tested the left-hand side of the screen against VidTest with similar results. But we had never tested the right-hand side.

Why? Because calling the same file, with the same calls, to the same lower-level programs, on the same machine would *obviously* produce the same result, right?

That fateful night, just for yucks, I tested on the *right*-hand side against VidTest. It was about 25% slower. Then I tested the right-hand side against the left-hand side. Again, 25% slower.

So, there we are. Same file. Same call. Same lower-level programs. Same machine. Different result.

I frantically began all the testing I should have done *before* shipping the product. We compared the display rate of all the codecs, in various resolutions and color depths, in both VCS display screens, against VidTest.

Here's what I found. The results were completely dependent upon resolution, color depth and codec. In some combinations, the results for both VCS screens and VidTest were identical. At other combinations, they differed by up to 25%. When they differed, the left-hand side of VCS equaled VidTest, and the right-hand side was slower.

I guess we all contemplate alternative careers at times like these. My favorite is gravel. I mean, what can go wrong with gravel? No bugs, no tech support, no languages to buy—you just carry it to some construction site, dump it, and you're done forever. But I was stuck with a bunch of newly minted CD-ROMs, no dump truck, and of course, a product about to be looked at by a reviewer who, as surely as day follows night, would discover this anomaly. OK, bug! So I needed an answer.

At 9:00 PM, I called Bubba, our engineer, at home. He, of course, thought my comments were ridiculous, denied it could happen, told me it was my machine, sun spots, local power problems, corrupted code, a funky hard drive, or bad karma, but in any case, certainly not his program. Not his baby. Having worked with engineers before, I told him to try it himself and call me back.

At 9:15 PM, at a vastly lower decibel level, Bubba called back. "What do you think it is?" he asked. Having worked with engineers before, I caught the "I told you so" before it got out, and replied, "I don't know."

But I passed along something I had noticed about VidTest. Between the time the video file opened, and when it began playing, the video window shifted one or two pixels in a seeming random direction. This is a bit hard to see in VidTest because it opens the file and then immediately plays it. You can see the effect more clearly by opening and playing a file with Media Player.

With VCS, the video display windows were embedded and didn't shift when the video started. The program is a 640x480 program that always loads in the upper left hand corner of the screen.

Bubba suggested that we move VCS's program window a few pixels to the right before playing a video to see if we could induce a shift. We tried it, and discovered that:

(a) we could induce a window shift by moving the VCS window to certain locations;

(b) when the VCS window shifted, the active display window performed identically to VidTest, whether it was the right or left window;

(c) once a shift occurred, the other display window played 25% slower than the first window that had caused the shift, and of course, VidTest; and

(d) once a shift occurred, the program window wouldn't shift again until you moved it to a new location.

So. We learned that a shifting window guaranteed the fastest video performance. We also learned that our dual-screen display, as currently configured, guaranteed that one window played about 25% faster than the other, irrespective of video or codec. Just the thing for a program billing itself as the ultimate video performance analysis tool, wouldn't you say?

10:30 PM. We took 10 minutes off to curse Microsoft, Video for Windows, the MCI specification, software, hardware, underwear and every other ware we could think of. Then we signed off for the night.

The next morning I called Steve the Video Genius, just to commiserate (I couldn't call him for help—his consulting rates are too high). He listened politely and said "Ya know, it sounds like the video is shifting to the 32-bit memory boundary in the video card. If so, data transfers to video memory are much faster—you send 32-bit blocks. If you're not on the memory boundary, you have to split the data into 16-bit blocks before transfer, which is less efficient and probably costs you a coupla frames per second."

"Oh," I said. "Sounds reasonable—let me check with Bubba."

Bubba had been busy perusing the Video For Windows SDK documentation for any information about the "shift." Predictably, there was none. Then he checked CompuServe, where he found some correspondence indicating that Steve the VG was correct. He also found some information indicating that the type of window determined whether the window would shift or not. This seemed like the true root of our problem.

Apparently, the MCI calls used in Video for Windows do not shift all types of video windows. Child and overlapped windows don't—pop-up windows do. So if you use a child or embedded video window in your application, it won't shift and you may not achieve the fastest possible display rate. This can cut your display rate by up to 25%.

Our product was a video performance analysis tool which we felt had to produce accurate results irrespective of window location, and—minor point—had to produce the same results on both sides. So we figured we had to use pop-up windows to force the shift.

Bubba estimated two days to turn the changes around. I called the reviewer, who said he wouldn't start looking at the program for a couple of days. Relieved, I replied, "Well, check with me before you start testing—we may have some interesting information."

So. We rewrote our code, changed our window style and everything worked fine, right? No, of course not. We're talking software here, and any time things can go wrong, they will.

Bubba turned the changes around in a record two days. I started testing, again around 7:00 PM. My wife, a first-year surgery intern, and I had our weekly two-hour staying-in-touch meeting scheduled for later that night. So I was hoping to quickly verify the results and then to be on my way.

This time, the right window was 50% slower. That's right 50%, as in five zero. Once again, same file. Same call. Same lower-level programs. Same machine. Different result.

Having come so far together, dear reader, I'm comfortable telling you that this was about my lowest point ever in software development. I felt the cloud of Murphy over my head, forecast loud and frequent ridicule in the trade press and an ultimate noisy and public financial ruin.

It was at this point that the phone rang. It was Bubba. "Hey Jan," he asked, "How's the testing coming?" I recounted my findings and he laughed, and said "Sorry, Jan, just a couple of missing pixels. Our video windows are too small. I was playing around at home and figured it out.

"When your video window is different than the actual video resolution," he continued, "Windows dynamically resizes the video to fit in the video window. We were forcing 320x240 videos to fit in a 318x238 box. This makes the computer do a lot of extra math to interpolate the video into the selected window. This costs you clock cycles and ultimately several frames per second.

"Don't worry," he concluded, "I adjusted the video window to a true 320x240, and now we're as fast as VidTest on both sides of VCS."

And that's how VCS Version 1.0 became Version 1.1. Our embedded windows are now perfectly sized pop-ups, and buddy, VidTest ain't got nothing on us from either side of the plate.

So, to recap. When using Video for Windows files, use pop-up style windows rather than embedded windows. If your windows don't shift, you're likely not getting the highest possible display rate. While this can be a bit of an appearance hassle, it's worth it in frames per second.

You might be thinking "I'll just line up the video on a 32-bit boundary and avoid the problem altogether." Well, this obviously can work, since the VCS' left-hand window was a child and its performance wasn't degraded—*so long as it was in its original position.* As soon as it shifted even one pixel, however, the video window was out of alignment and performance suffered.

And there's always this enigma—both VCS display windows were on a 32-bit boundary, yet performed differently. Each window was 320 pixels wide, which is 80 32-bit, or 4-byte, memory blocks. So if the right window was aligned properly, the left window started exactly 80 blocks over and was also aligned properly. If both windows were on a 32-bit boundary, why did performance differ? So it's probably safer to use pop-ups in all situations.

The second point is that irrespective of window type, make sure your window size exactly matches your video resolution. Otherwise, you may be forcing Windows to dynamically resize the video at a price of several frames per second.

This can be pretty tough, because many authoring programs force you to draw the window, not create it to a specific pixel size. For example, ToolBook 1.52, which we used to develop the multimedia *Guide to Video Compression*, forces you to draw the window by hand (Version 3.0, which shipped this summer, lets you specify exact pixel size).

To make sure we had the exact window size, we ported VCS's display rate DLL to calculate the display rate of our embedded video from *inside* the application. By comparing this to the results achieved in VCS, we make certain our window size is correct. It took some tuning, but now we know we're getting the most from our video.

Oh—by the way. The reviewer who started this all—didn't look at the product. Never cracked the shrink wrap. I guess that's life in the fast lane. His loss, right? He missed a great story.

YOUR CAPTURE STATION

14

YOUR CAPTURE STATION

This chapter will describe the hardware and software requirements for your primary capture and compression station. On the hardware side, we'll look at components to spec into your new system, and how to beef up your old system for enhanced performance. On the software side, we'll look at programs you'll need for capture, preprocessing and final compression.

We will not look at CD-recordable systems, that being a topic worthy of separate consideration. For similar reasons, we will not address morphing, image editing or authoring programs.

Processor

As I sit here and read advertisements in the August, 1994, issue of *PC Magazine,* the price difference between a 486/66 VLB computer and a comparably equipped Pentium/90/PCI hovers at around $500 for Ambra Computers (RIP) ($2,499 vs. $2,999), $700 for Gateway Computers ($2,299 vs. $2,999), $900 for Micron Computers ($1,899 vs. $2,799) and $650 for Zeos Computers ($2,095 to $2,745). And *PC Week* recently reported that Intel was dropping Pentium prices rapidly to slow the sales of 486 clones by AMD, Cyrix and other competitors, so prices will only be dropping.

Let's say your CD-ROM title has 60 minutes of video. It takes close to two hours to compress one minute of either Cinepak or Indeo on a 486/66 computer, so that's about 120 hours of compression. It takes less than one hour to compress the same minute of video on a Pentium 90.

On final compression alone, you'd save about 60 hours of compression time. Throw in prototype compressions, editing, filtering and assorted other time-consuming but very necessary video production tasks, and the Pentium that cost you less than $600 more saves you about 100 hours of time. That's about $6.00 per hour.

Bus

None of the four companies mentioned above offer a Pentium computer with VESA local bus graphics. They're all PCI. Three of the four offer PCI-based 486 computers.

However, don't shut off brain once you see the magic PCI acronym. Here's a couple of other points to consider.

The first local bus machines were all "local bus graphics" machines. The local bus accelerated the transfer from main memory to the video card, but little else.

As we saw when looking at the capture process, transfer speed from main memory to the hard disk is critical to overall capture performance. Given the history of local bus computers, you should always determine whether the connection to the hard drive is PCI or ISA.

Fortunately, the clear trend is towards local bus hard drive connections. For example, AMBRA Computers advertises a "Fast PCI/IDE Controller," Gateway a "PCI Enhanced IDE Interface," Micron a "9 ms SCSI-2 hard drive (PCI)," and ZEOS offers a "local bus IDE hard drive." If you're buying a second- or third-tier brand, or assembling your own computer, make sure that the drives sit on the PCI bus.

The second point is slots. You're going to need lots of them. And they're going to have to be PCI slots, not ISA, or you won't get the benefits you're looking for.

Don't buy a three-slot machine like the IBM ValuePoint, even though the slots are VESA/PCI. Period. These are great end-user machines, but you'll need more slots to produce video.

Here are the connections that will have to be local bus.

1. *Graphics Adapter*—We'll chat in a minute about my bias against embedded graphics adapters. Even if your computer offers video on the motherboard, you'll have to try all sorts of new graphics things, such as video accelerators, ZOOMDACs, and 64-bit/128-bit graphics adapters. So ultimately you'll need a free PCI slot for video.
2. *Drives*—We're going to recommend that you have at least 2 gigabytes of fixed storage. Most computers come with one fixed disk, which you'll probably need to supplement with a second. At least one, and preferably both need local bus connections. In theory, one SCSI controller can control up to seven devices, but I don't know if or how much this will affect performance during capture.
3. *Capture Card*—While only one PCI-based capture board is available as of this writing, they're coming, and if you buy a PCI machine now you'll want a PCI capture card now or later. And, of course, you'll need a slot to put it in.

Some combination of available PCI slots or embedded local bus drive controllers should support all these peripherals. Other peripherals such as modems, network cards, sound boards and CD-ROM controllers don't need PCI-level throughput and can occupy ISA slots.

RAM

The amount of installed RAM ranks third behind processor and bus in terms of boosting overall system performance. Many video tasks are extremely input/output-oriented, and the more video you can store in main memory, the less the processor has to page back and forth to disk to get the job done.

In addition, more RAM also translates to more video captured without compression. With 64 MB of RAM, for example, you can capture close to 60 seconds of 320x240 15fps in YUV mode without compression. This means better-quality video in bite-sized chunks you can easily paste together for longer clips.

Capture aside, however, you'll probably experience your biggest benefit jumping from 8 MB to 16 MB of RAM, with diminishing returns thereafter.

Graphics

Twenty-four is the magic number. The eye can perceive over 16.7 million colors, and you need every one of them to make video look lifelike. As we learned, 24-bit video doesn't cost you storage space or bandwidth because all the high-end codecs store information in 24-bit format. However, the 24-bit codecs dither in 8-bit environments.

You would think that you could assume 24-bit graphics with high-end computers, but you really can't. I recently taught a two-day seminar for SIGCAT in Reston, Virginia. They were kind enough to send down a computer to load and configure in our offices and then use for the class. Saves wear and tear on our trusty development Gateway.

They sent down a Compaq ProLinea 486/66 mHz VESA local bus computer. I loaded up Video for Windows and VCS and played the first video. Dither city. No problem, I thought, I'll just load the 24-bit drivers. Click, click into Windows Setup where I discovered no 24-bit drivers. No 16-bit drivers. Only 8-bit. I checked the manual and 8-bit was all the bit we were going to get.

No problem, I thought. I'll just install the VideoLogic 928, a VESA local bus card instead. I opened the computer and discovered a total lack of VESA slots. Zero. Nada. Nein. We not only couldn't upgrade the graphics on the motherboard, we couldn't install an alternative video card.

Segue to the First Looks section of the August *PC Magazine*. The Compaq Deskpro XL reviewed on page 42 offers

"1,280-by-1,024 resolution with 256 colors and 2 MB of VRAM." A $3,299 Pentium computer (monitor extra) with 8-bit graphics.

Now this doesn't mean that higher color depths aren't supported at lower resolutions—but the article didn't mention that they were. It just means that even with high-end computers like the Compaq, you've got to ask the questions.

Monitor size is somewhat less relevant. While bigger is definitely better, anything in the 15–16" range is acceptable.

Fixed Storage

Your capture system should have at least two gigabytes of on-line storage on your capture system, with additional storage accessible through an off-line system like a tape drive. You'll be working with capture files that can easily exceed 200 megabytes, and filter and edit files that are just as large. As you develop your project, product or presentation, you'll accumulate bulky finished video files along the way.

Consider additional space if you plan to create one-write CD-ROMs from the capture system with a CD-recordable unit. The premastering software used by one-write systems typically requires you to build a mirror image of the target CD-ROM on your hard drive. In other words, if the content in your CD-ROM is 650 MB, the software will need another 650 MB to build the files and file structure to be burned on the CD-ROM.

Mirroring accomplishes two things. First, with many premastering programs, it lets you test the integrity of the CD-ROM. For example, you can test links to video and other files, debug program software and in some instances test performance with so called CD-ROM emulation software.

The mirror image also ensures that the data feed to the one-write system will be smooth and unbroken, since the hard drive doesn't have to search during the process; it just reads the entire mirror image from start to finish. Any break in the data feed to the write systems typically ruins the one-write media, creating what's called a "coaster," or a $15.00 gold

CD-ROM suitable only for protecting your fine office and lab furniture.

There are premastering programs that don't require a mirror image and build a "look-up table" for the included files instead. I haven't tested programs that don't mirror, but with an overall success rate of around 60% for our one-write efforts, I'm not about to try anything that could possibly make it worse.

The type of drive that you purchase is getting rather intriguing. Until recently, you could purchase any old drive so long as it had a fairly fast seek time—typically you'd try to get under 10 milliseconds. However, Micropolis just introduced a line of "AV" drives optimized for video and audio capture and playback that could change the equation. Here's the story.

Hard disks perform an array of housekeeping functions, such as rotational retries, thermal recalibration and data head degaussing, to reliably store and retrieve data. During these housekeeping functions, reading and writing to and from the disk is stopped.

Most drives achieve their rated throughput, usually measured in megabytes per second, through a combination of abrupt housekeeping stops and burst modes. For example, a three megabyte per second drive might read/write at 3.15 megabytes per second for 19 seconds and then stop for one second for housekeeping. Overall this computes to three megabytes per second. However, during the one-second stoppage, no data can be written or read, creating dropped frames during capture and jerky video during playback.

For this reason, for video capture and playback, average throughput isn't relevant. Rather, it's the drive's ability to sustain a read or write without interruption, essentially delaying housekeeping functions for mission-critical input/output.

Let's look at thermal recalibration as an example. Thermal recalibration is defined by Micropolis as:

> A method used by hard disk drives to automatically compensate for expansion and contraction of mechanical components caused by temperature changes. Typical drives perform thermal recalibration at regular intervals. During the

thermal recalibration process, reading or writing of data is suspended, possibly interfering with real-time data transfers.

Apparently, as hard drives heat up during normal operation, the disks expand and the disk heads must adjust to remain synched with the data tracks on the drive. This adjustment, called thermal recalibration, can take up to one-tenth of a second, during which time no information can be written to or read from the disk.

Micropolis' AV drives prioritize user read/write requests over regularly scheduled thermal recalibrations. The disk won't interrupt user operations to recalibrate, as other drives will, and will interrupt a recalibration to support a user I/O request. This should translate to fewer dropped frames and smoother playback.

Other housekeeping activities are similarly managed, resulting in smooth data input/output to and from the disk drive. This could enhance real-time capture and video playback, especially if you're working with capture data rates in the megabytes per second.

Digital Video Magazine reviewed a Micropolis AV drive in its September, 1994 issue, running the drive overnight with no evidence of thermal recalibration problems. The magazine stated that the drive was "a great value and an excellent choice."

Other manufacturers of drives that do not perform thermal recalibration are Seagate (the Elite and Wren models), Digital Equipment Corporation (DSP3105) and Quantum (LPS540 and Empire 1080S).

Single or Multiple Drives

Your total drive space can be allocated to one drive, or multiple drives. While it's nice to purchase your system entirely pre-configured from the manufacturer, which typically means one large drive, something about having your entire system on one drive is scary. Our system has the 350 megabyte drive shipped from Gateway, and another 1.4 gigabyte drive we subsequently installed. The second drive crashed once and we were able to operate the computer normally using only the smaller drive.

Off-Line Storage

I don't have a lot of value to add on off-line fixed storage, having limited experience with tape drives and other similar products. We have used our CD-recordable system for backup with poor results. The inevitable coasters (failed discs) get expensive, the software is not designed for general backup use and is difficult to use in that role, and you have to clear space for a mirror image, which you don't have to do with other methods. I recommend picking a common tape format.

If you want to use your CD-R system as your primary backup mechanism, make sure the software has a designated backup mode. Our software won't write an empty subdirectory onto the one-write CD-ROM, which means you can't wildcard transfer all of your subdirectories. While this feature is of obvious benefit when writing a product-bound CD-ROM, it's a real pain in the rear when backing up, because you have to list and check every subdirectory.

In addition, make sure you can deal with the space required for mirroring. If you've got 1.7 GB of disk crammed with finished video files and have to mirror, you've painted yourself into a really tight corner. Oftentimes a cheap $250 tape backup system will give you acceptable results at reasonable prices.

SCSI vs. IDE Controllers

There are two kinds of disk controllers, SCSI and IDE. IDE was the first controller standard but lost momentum because it couldn't handle drives larger than around 500 megabytes as a single logical unit. SCSI, which stands for Small Computer System Interface, assumed IDE's lost momentum by addressing unlimited disk sizes, with the added benefit of being able to manage up to seven peripherals through one controller.

The new Enhanced IDE specification overcomes the fixed disk size limitation and allows compatible boards to address disks as large as SCSI controllers—all the way up to 8.4 gigabytes. One controller can also drive up to four IDE devices, which includes hard drives, CD-ROMs and tape drives. Some new IDE controllers also have SCSI connectors.

Not to be outdone, SCSI responded with SCSI-2, which increased the bandwidth and speed of SCSI transfers. Nonetheless, much of SCSI's perceived advantage is gone, at least from a pure disk-performance standpoint.

Overall, having at least one SCSI drive on your computer adds a lot of input/output flexibility. However, many new computer systems ship with IDE drives, which really shouldn't disqualify them from consideration. Just make sure that you have one free PCI slot for the SCSI card, or you won't have local bus transfer to and from the drive.

NTSC Monitor

First on our product wish list is an NTSC monitor, which is essentially a television without separate tuning equipment. Instead of being driven by signals received from the airwaves, it's driven directly by your analog source.

Most analog decks have at least two outputs, usually composite and S-Video. This lets you connect the S-Video to your capture card and the composite to the NTSC monitor. When you play the video, it outputs through both cables to the capture card and the monitor.

The helps in several ways. First, when installing your capture card, if video isn't present in the capture window, you have no way on knowing whether the deck is outputting video. If the monitor shows video, you know that's not the problem and can focus your efforts elsewhere.

Second, during capture, many boards freeze the capture window to focus their efforts on capturing and storing data. This means that the video window doesn't update along with the video, and you can't see when to stop the capture. Your only alternative is to time the capture, which is pretty inexact.

If you have an analog monitor, you can watch in real time and stop the capture when appropriate.

If a portion of your capture footage is blotchy or blurred, you can check the NTSC monitor and see if the errors relate to the original footage or your capture system. It also shows the color and brightness of the original analog footage so you can match the digital video more closely. Step by step through the capture process, the analog monitor provides valuable system feedback, saving time and improving overall results.

You can purchase an NTSC monitor at your local Audio/Video store. Prices start in the $200–300 range and increase by monitor size. While many low-end monitors are gray-scale, color is preferable for synching digital and analog color and brightness. Alternatively, you can use any cable-ready color television. It won't look as cool, but it will get the job done.

CD-ROM

A double-speed CD-ROM is nice for quick loading and reloading of the many programs you'll be using. Just be sure to keep a single-speed drive installed somewhere in your operation for testing purposes.

BEEFING UP THE OLD GRAY MARE

If you've decided to use a 486/66 that you already own, adding additional RAM will provide a quick performance boost. In particular, boosting RAM from 8 MB to 16 MB makes a world of difference.

You also might want to consider adding a local bus hard-disk controller to your system, which should increase capture, playback and editing performance immensely. Available in both Enhanced IDE and SCSI-2, these controllers obviously require an open VESA slot. In return, they deliver roughly 10 times the performance of ISA controller cards, with burst modes up to 11–13 MB/second. These products are now available from Adaptec, Acculogic, GSI and others.

SOFTWARE REQUIREMENTS

Video for Windows Tools

Microsoft has reportedly decided to remove the Video for Windows (Fig. 14.1) application suite from the market. This means that you won't be able to purchase VidEdit, VidCap, VidTest, *et al.* from Microsoft in the future. It would appear, however, that while these specialized tools may disappear from the market, the underlying Video for Windows architecture will live on.

While your capture card will undoubtedly ship with tools that provide similar functions, at certain tasks VidEdit and VidCap excel in both performance and strict ease-of-use. VidEdit also provides certain unique functions that you simply can't get anywhere else, such as computing the optimal palette for a group of disparate images.

Finally, as we discussed, few programs have been able to exactly emulate VidEdit's compression performance. You'll always want a copy of VidEdit around, if only to compare results with those achieved with your compression program.

Figure 14.1 Video for Windows application. Endangered species?

If you have these tools, don't throw the disks away. You can acquire the tools on the Multimedia Jump-start CD-ROM, which Microsoft distributes free. Fax your request for the Jump-start disk to their Multimedia Developer Relations Group at (206) 936–7329. Alternatively, you can request the disk on the Microsoft Video for Windows forum in CompuServe. I don't know how much longer Microsoft will continue to offer the Jump-start disk, but it's worth a phone call.

If you can't get the Jump-start CD-ROM, you might hunt the computer trade magazines for older boards that shipped with Video for Windows. The original Intel Smart Video Recorder did, as did Media Vision's ProMovie Spectrum. Both boards should be available for under $100 by the time you read this book—throw away the board, and keep the software.

Video Editing Software

You'll need software to perform titling, transitions and other functions not performed by VidEdit. As of August 1994, Adobe Premiere has achieved a dominant position in this arena, having an extremely strong product entry on the Mac side and an evolving product on the Windows side. However, the market is new and very dynamic. Asymetrix's Digital Video Producer, included with Smart Video Recorder Pro, should provide competition, as should U-Lead's Video Studio.

In truth, you'll probably end up using whichever editing program comes with your capture board. If you acquire a capture board without editing software, I'd recommend Premiere for industrial-strength video editing, and Digital Video Producer for ease of use and good performance on a moderate systems.

Audio Editing Software

Your capture card should ship with software that enables rudimentary cutting and pasting. I've worked with Media Vision's and Creative Labs', and both were sufficient for the limited type of audio editing that we perform at Doceo.

If you need advanced capabilities, track down the latest sound board and audio software review in *Morph's Outpost, Multimedia World* and *New Media Magazine,* who all do a great job following these product categories.

The Norton Utilities

Video capture, editing and compression can get your hard drives out of joint in a hurry. Rarely is there a day where Norton doesn't find a misallocated temp file or some other disk error on one of our two systems. Cleaning these up in real-time avoids accumulations that can lead to hard drive crashes.

In addition, Norton's tools are more robust than those available in other programs, even those licensed from Norton! At a recent seminar, I was transferring files with LapLink from one system to another, and the file and directory structure on the target system got completely corrupted. The files and subdirectories wouldn't delete using any DOS and Windows command known to man. Then we tried the defrag disk utility licensed from Norton that's included in DOS 6.2. No help.

The technician I was working with suggested Norton Disk Doctor. I scoffed, because I knew that the Microsoft tools we had just tried were licensed from Norton. But we tried it anyway, and Norton fixed the problem.

The Norton Utilities are owned by Symantec Corp., who can be reached at (800) 441–7234.

Video Compression Sampler

You've spent some time with VCS Play (Fig. 14.2), and I hope you've come to like it. The retail version loads without the CD-ROM in the drive, and also includes benchmark files for all the codecs discussed in this book. We also issue periodic updates with new codecs.

Figure 14.2 VCS—the program that launched this book

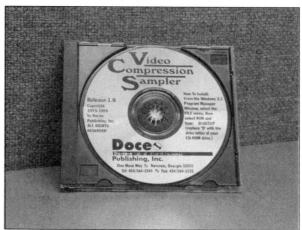

You can upgrade to the retail version for $50, a savings of $50 off the retail price. Call Doceo Publishing at (404) 564–5545 to order.

Doceo Video Filter

While most appropriate for talking-head and other low-motion videos, this program is worth its weight in bandwidth reductions and overall higher-quality video. If you plan on using the Indeo 3.1 Quick Compressor or Indeo 3.2, our Filter is especially helpful. The program retails for $299 and is available directly from Doceo Publishing.

Doceo Batch Compressor

When we launched this product, the press release headline read "Doceo Announces Batch Compressor—Pizza Stocks Tumble." Compression, being the final stage, occurs last, usually when the product is long overdue. This means night and weekend work, and pizza for the operator who has to load a new video file to compress every hour or so.

The Doceo Batch Compressor duplicates VidEdit's compression functionality in batch mode, letting you compress in unattended mode at night and on weekends. As of the time of this writing, it is the only batch compressor available. Save yourself time and heartburn. The program retails for $299 and is available directly from Doceo Publishing.

SUMMARY

1. If you're buying a new computer system, here's what to look for:

   (a) *Processor*—Pentium/90

   (b) *Bus*—PCI, with enough slots for hard drive, video graphics and capture card

   (c) *RAM*—At least 16 MB

   (d) *Fixed Storage*—At least 2 gigabytes, sitting on the PCI bus. Check out the new line of AV drives from Micropolis.

   (e) *Off-line Storage*—CD-recordable units are discouraged. Get a tape or similar system.

   (f) *SCSI vs. IDE*—Performance is very similar, but SCSI-2 controls a wider variety of external peripherals. Make sure the controller sits on the PCI bus.

   (g) *NTSC Monitor*—Any one will do.

   (h) *CD-ROM*—Double speed.

2. Two cheap and dirty ways to boost performance of your current system:

   (a) Add RAM (at least to 16 MB).

   (b) Add a local bus disk controller.

3. Software you can't live without:

 (a) *Video for Windows Application Suite* (VidEdit/VidCap/VidTest/ScreenCAP).

 (b) *Video Editing Software*—Premiere if you're buying.

 (c) *Audio Editing Software*—Should come with your sound board. If not, check the latest *Morph's Outpost* or *New Media Magazine*.

 (d) *The Norton Utilities*—Lock the barn door before your workhorse crashes.

 (e) *The Video Compression Sampler*—Of course!

 (f) *Doceo Filter*—Primarily for low-motion sequences and Indeo 3.1 Quick Compressor.

 (g) *Batch Compressor*—You deserve the night off! Buy the Batch Compressor.

MPEG

15

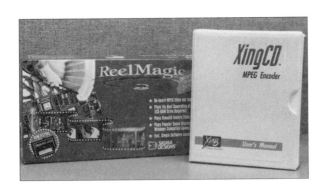

MPEG

MPEG is like a technology and a revolution. As a technology, it offers 30 frame per second performance at close to VCR-like quality. The obvious downside is that you need a decompression board in every target computer.

As a revolution, MPEG has gained substantial momentum over the last few months, starting with the release of Sigma Designs' RealMagic board. Initially, the market and trade press scoffed. But then companies such as Creative Labs and Matrox introduced MPEG Playback boards, and Texas Instruments and IBM announced MPEG playback chip sets. Suddenly, MPEG was everywhere, seemingly peaking around the May 1994 COMDEX in Atlanta.

This chapter will examine the technology, in the form of two entry-level MPEG products, Xing CD, a software-only MPEG encoder, and the RealMagic playback board. Then we'll stare into our crystal ball and no doubt make some predictions we'll regret by the time this book hits the streets. Ah, the computer business.

HISTORICAL PERSPECTIVE

MPEG-1

MPEG stands for Motion Picture Experts Group. The first MPEG Standard, known as MPEG-1, was introduced by the committee in 1991. Both video and audio standards were set, the video standard built around the Standard Image Format (SIF) of 352x240 at 30 frame per second at a data rate of 1.5 megabits per second, or about 170 kB/second. For obvious reasons, most MPEG-1 products are compressed at about 150 kB/second.

As we saw in Chapter 2, MPEG uses a combination of inter-frame and intraframe compression. Where Video for Windows uses key and delta frames, MPEG uses "I" frames, their analogue to key frames, and "B" and "P" frames, which are delta variants. B frames are bidirectional and can acquire data from I frames located before and after them in the video stream. P frames, or Predictive frames, are delta frames that only derive data from I frames that precede them in the video stream. Not surprisingly, MPEG's intraframe is JPEG, the still-image standard spawned by the Joint Photographers Experts Group.

By resolution and data rate, MPEG is targeted primarily at the computer and games market, as opposed to broadcast quality. Video quality is most frequently compared to VCR quality.

MPEG-1 immediately spawned no fewer than five derivative standards, called Yellow Book, Green Book, White Book, Red Book, CD-I, CD-I and Video CD. Once, on a flight from Atlanta to New York, I devoted two hours to deciphering the different standards. I failed miserably, despite having only one Scotch.

This alone convinced me that without outside intervention, MPEG-1 would never make it as a computer standard. After all, if a reasonably intelligent person totally immersed in video can't tell these names and colors apart, how can the average developer, VAR or MIS director?

To make matters worse, vendors such as Sigma Designs attempted to lock in applications developers by keeping their APIs proprietary and out of the hands of other hardware developers. Title developers were forced to develop for one MPEG playback board or another, which fractured an already tiny market and created another standards vacuum that was quickly filled by Big Brother from Redmond.

Microsoft

On June 2, 1994, Microsoft released Version .99 of their MPEG Command Set for MCI. As with Video for Windows, the specification provides an inbound architecture for MPEG hardware and software developers, and an outbound architecture for applications developers seeking to use MPEG in their products.

Overall, this specification will help standardize development in the MPEG community, which will certainly promote the MPEG standard. This, combined with a DCI-like standard for MPEG in Chicago illustrates, in their own words, that Microsoft "is serious about making Windows the best platform for multimedia development" (from the CompuServe announcement).

Interestingly, the Microsoft announcement came less than a month after Apple's announcement that MPEG would be supported directly in QuickTime. That hearkens back to Microsoft's original Video for Windows announcement that came in a rush the same day that Apple announced QuickTime for Windows.

MPEG-2

MPEG-2, adopted in the spring of 1994, is a broadcast standard specifying 720x480 playback at 60 fields per second at data rates ranging from 500 kB/second to over 2 megabytes per second. At these rates, MPEG-2 is not particularly relevant to the computer industry in the short term.

What's interesting about MPEG-2 is a total wrangling over patents and royalty rights by the committee members. MPEG-1 is an open standard that any hardware or software vendor can implement. Many committee members who contributed technology to MPEG-1 feel the need to cash in on MPEG-2, which caused serious delay in the adoption of the standard.

Whether this scares potential MPEG-1 adopters away remains to be seen. On the other hand, it clearly negates much of the certainty usually derived from working with a "standard," such as a known cost structure and confidence about the technology's future.

THE PRODUCTS

XingCD

XingCD, from Xing Technologies, is a software program that takes in digitized video and spits out MPEG files. The two-step process is very similar in operation to any Video for Windows codecs—you just use a different set of tools and get a file with a different last name.

The product costs $995. In contrast, real-time capture systems start at $5,000 and quickly jump to $20,000 and higher.

Figure 15.1 XingCD Encoder screen

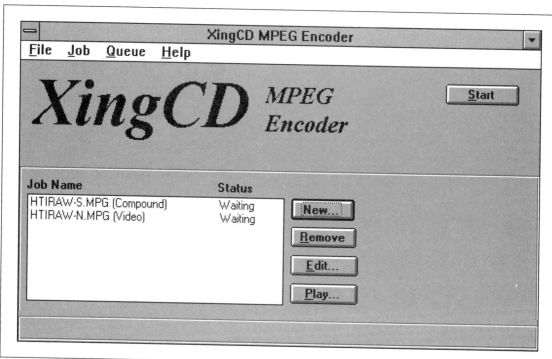

Figure 15.2 The curiously dithered look of XingCD in software-only playback

The product consists of an encoder (Fig. 15.1) and a player. The encoder, essentially an MPEG VidEdit, accepts AVI files, streams of Targa and 24-bit BMP files and other MPEG files and outputs three flavors of MPEG, White Book, for playback on the Philips Video-CD, XingCD format and a quarter-screen MPEG format called QSIF. We compressed all our files to XingCD format. Compression took about an hour per minute of video, about the same as Cinepak or Indeo. The program includes a batch mode that worked very well.

We couldn't use Video for Windows tools such as VCS Play to measure final file bandwidth. We eyeballed the files with File Manager and looked close to the target data rate.

As expected, playback in software was extremely slow. Xing's Player software lets you play back in two modes, full speed, where it drops frames to keep up with the audio, and No Rate Control, where it plays back every frame. Playback of a 20-second file in the second mode took 76 seconds on our 80486/66, or about 8 frames per second.

When playing back in software, Xing's files all looked extremely dithered (Fig. 15.2), almost distractingly so. On the RealMagic hardware, the files were superior to all software codecs, but not perfect. The Action sequence exhibited some of the same jaggy-edged motion artifacts as Cinepak, and the JPEG artifact "Gibbs" effect around the numbers.

We didn't compare our files to the output of any other encoders. However, the video quality produced by XingCD was subjectively very high, and the program installed perfectly and worked well throughout. Overall, if you're interested in testing how MPEG will look on your files, XingCD would be a great way to go.

RealMagic

We expected that MPEG's quality would exceed that produced by Video for Windows, and we were right. The only question was, what did 30 fps performance cost you? For that answer, we move to the RealMagic Board (Fig. 15.3).

Installing the RealMagic board was a pleasant surprise. It takes a 16-bit slot and requires that you connect an overlay cable to the video card. After what we've been through with some of the capture cards, however, it was a breeze.

When we looked at the board's I/O requirements, we thought that software installation would be extremely challenging. Depending upon your configuration, the board can require up to three I/O ports, one DMA channel and an interrupt. However, the board's DOS-based installation software located

Figure 15.3 The RealMagic Board and XingCD

open settings and included diagnostic routines to debug installation or configuration problems. Sigma even included an uninstall program.

Problems started in Windows, as happens with many overlay cards. Program Manager became almost totally black and the icons and text got extremely distorted. The trouble-shooting guide recommended running in 256 colors, so we shifted down to 8-bit mode. Didn't help. Finally, in 16-color VGA mode, the system stabilized and we were able to run the videos created by XingCD.

Which looked great. The dithered look of the software-only playback totally disappeared. The videos were smooth, clear and played back without a hint of interruption. Unfortunately, we can't show you any screen shots. As an overlay card, RealMagic card doesn't work through Windows; it decompresses the video on board and shoots a signal to the screen directly through your video card. You can't do a Windows screen capture—there's no screen. So you'll just have to take my word.

Overall, the concept of video overlay tended to dominate your environment, not smoothly fit it. In the context of a kiosk, or encompassing game, the technology felt right. But in the context of your normal multitasking environment, video overlay didn't fit in. Throw in the fact that all applications were limited to VGA color depth, and I was happy to use the uninstall routine and get the board out of my computer.

The next generation of MPEG Playback boards from Matrox, Jazz Multimedia and others integrates MPEG playback with normal video graphics functions, including video chip sets capable of 24-bit playback. The card is both graphics card and MPEG Player. The Matrox card looks particularly interesting, integrating video co-processor functions with the MPEG Playback.

Maybe these products will integrate into the business environment more smoothly than the RealMagic card and change my opinion. For now, however, the current MPEG technology feels application-specific, used best in targeted applications like kiosks and games.

RealMagic, from Sigma Designs, retails for $349. You can reach Sigma at (510) 770–0100.

THE TECHNOLOGY

Winding the String

Let's look at MPEG from a pure technology perspective. We'll start with a question.

Which is more technologically advanced, MPEG or Indeo? MPEG can compress 30 fps down to 150 kB/second and play back in real time with a $300 decoder board. Quality is very good, but not exceptional.

Indeo can compress 15 frames per second down to 150 kB/second, and play back in real time in most 486-class local bus computers. Quality is good. At 30 frames per second, quality is poor.

So. Which is more technologically advanced?

At the start of this book, we saw that codecs have a three-part challenge, and must juggle both video quality and display rate while meeting the target bandwidth. Software codecs walk a very narrow road. Compress too much, and they lose quality. Use innovative techniques such as motion compensation, and risk display rate.

In many ways, compression is like winding a ball of string. Take a long time and you can wrap it very tightly. However, it may be very difficult to unwrap.

But suppose you knew you had a special "string unwrapper chip" at the other end. You could wind really tightly because you'd know that when the time came, you could hand the ball off to this special chip and it would unwind the string at 30 fps.

How much of MPEG's compression efficiency relates to the chip at the other end? Let's do an experiment. We'll take two compressed files, an Indeo and an MPEG file, and use PKZIP, probably the world's oldest compression technology, and see how much each file compresses (Table 15.1).

As you can see, the MPEG file only compressed 4%, while Indeo compressed an additional 30%. Which means one of two things. Either the Indeo engineers weren't familiar with the finer points of lossless compression, or they couldn't wind the string any further and still display at 15 frames per second.

Remember that Indeo came from DVI, a hardware-based technology that rivaled MPEG in video quality. It's relatively clear that Intel spent 1992 and 1993 defeaturing DVI so it could play back in software. Now, with Indeo 3.2, they're adding features back, and quality and playback rate are improving.

Table 15.1 Effect of PKZIP on MPEG and Indeo files

Codec	Original	Zipped	%
Indeo	2,022,504	1,436,574	30%
MPEG	1,947,215	1,880,153	4%

Today, both Indeo and Cinepak can playback 30 frames per second on our trusty 486/66. MPEG gets about eight.

MPEG is based on JPEG, a still-image technology that's pretty much maximized its potential. Indeo and Cinepak are based on vector quantization, a technology with plenty of headroom for further advancement.

MPEG was designed in 1991, when the 386 was state-of-the-art and the local bus was one that stopped at every street. Today, with video co-processors, DCI, local bus and Pentiums being the norm, the power behind the video rivals that of an MPEG decompression board. If Intel and SuperMatch tuned their algorithms to perform only on higher-end computers, they, too, would deliver MPEG like performance.

So which is more technologically advanced, MPEG or our software codecs? From where I sit, it's the software. MPEG's performance relates primarily to the decompression hardware. Take the hardware away, and it's unmarketable.

MPEG can't claim technological superiority. In a few months, they may not be able to claim performance superiority.

The Concept of Usable Video Quality

I begin all seminars with the concept of usable quality. There's always one individual who says that 30 frames per second at full screen is absolutely essential. Will accept no less. I begin to sweat slightly and wonder why this person paid hundreds of dollars to learn how to use a technology that by definition doesn't meet his or her requirements.

Then we discuss Sherlock Holmes, which sold over 400,000 copies with video inferior to Video 1. Or the Seventh Guest, with sales in seven figures, that features gauzy, ghosty video. I guess it works because it's a haunted house. We talk about

Nintendo and Sega, products that sold in the millions with quality vastly inferior to the average Saturday morning cartoon. Clearly, from an entertainment standpoint, broadcast quality isn't necessary.

We move to the concept of video as a medium for information transfer, and watch the short talking head video. Is this more effective than a written help file? Everyone agrees.

We review the short installation clip showing the lineman suiting up for pole climbing. Is this more effective than text or even graphics? Everyone agrees.

We discuss all the uses of video, and pretty much everyone agrees that 30 fps, full screen quality isn't essential. Preferable, maybe. But not essential.

The key point is this. Television quality isn't absolutely necessary for most uses of video—especially business uses. A properly prepared, professional presentation can be just as effective as broadcast quality, and much better than most alternatives.

CRYSTAL BALL

So here's my crystal ball.

General Corporate

The installed base, general-use computers—No way will even a small fraction of these machines be upgraded to hardware MPEGs. Too expensive, and video quality isn't necessary for the essential purpose of corporate video. Most computer users won't give up 24-bit graphics for the occasional video.

New Computers

On computer motherboards—No way. Compaq still uses 8-bit graphics on most machines; you think it will rush to adopt MPEG? I don't think so. Standards change too fast, and motherboards with dedicated chips are obsolete inventory waiting to happen.

On standard issue graphics cards—Tougher call. DCI support is a given, and I'm betting that video co-processors will become pervasive. Most corporations will not pay extra for MPEG support. I don't see MPEG as standard issue on these cards for at least a year, and maybe never.

Training, presentations and kiosks—Absolutely. Limited-distribution applications are perfect for MPEG, after you get over the hump of integrating MPEG into your development environment.

Video conferencing—No way. The cheapest real-time MPEG encoder is $5,000, while Smart Video Recorders, the compression engine for Intel's video conferencing system, costs $500. Video conferencing is probably Indeo's largest raison d'être.

The theory of convergence sees video lying between three markets—computers, entertainment and telephony. With videoconferencing touted as the primary corporate uses of desktop video in the '90s and beyond, it's interesting to consider how MPEG's inability to play in this market will affect its penetration of other markets.

Home

Cable set-top boxes—Absolutely, but probably MPEG-2. This market requires at least VHS quality.

I don't see MPEG-1 movies on CD-ROM being a strong market. The parallel to CD-ROMs and records simply doesn't work—CD-ROMs sounded better and were easier to maintain than the vinyl equivalent. In contrast, MPEG-1 can't match VHS quality, and there are probably close to 100 million VCRs in homes today. At today's CD-ROM densities, most movies would require at least two CD-ROMs. You also can't shoot home movies, pop the tape and play it on your CD-ROM.

Computer add-in boards—Possible, but not likely. *Business Week* recently reported that the home market was grabbing an incredibly high share of all Pentiums sold. All are local bus computers that play 320x240 video with ease. These are not good candidates for MPEG upgrade cards. But then again, I don't have a six-year-old or a teenager.

Computer games—As in Atari, Nintendo and 3DO. Absolutely. This is probably going to be MPEG's largest market in the short term.

SUMMARY

In the country of the blind, the one-eyed man is king. In the country of the software codec, the hardware-based technology is king. But only so long as the underlying hardware doesn't catch up. MPEG can wind the compression string tighter, but only because of its decompression hardware.

In 1991, when MPEG was announced, the average computer was a 386 with ISA bus. Software codecs were crippled by processing power, bandwidth limitations and Windows.

By June 1995, the average corporate entry-level computer will be a Pentium/100, more than enough to play back 30 frames per second of either Indeo or Cinepak. The bus will be PCI, with more than enough throughput to support this level of playback. The Display Control Interface will be standard, and Windows will no longer drag the performance of software codecs. Video co-processing chips will be on every graphics card, enabling CPU-free scaling to full screen.

In the great "who cares the most" analysis, no big players have a real stake in MPEG. Committee members are now wrangling over percentage, not market strategy. The entertainment community doesn't care which standard—they just want reliable broadcast quality. Ditto for the home consumer—they just want their MTV. And if you added up all the MPEG hardware sold last year, it probably wouldn't equal one day of Intel's sales.

Hardware standards must achieve critical mass before the power in the host CPU renders them useless. Otherwise, they get relegated to niche markets. By mid-1995, video co-processors, local bus technologies and DCI will provide a hardware platform to boost software-only technologies far beyond MPEG.

So don't get caught up in the MPEG hoopla. It's bound to have its markets, but it's surely not the last word in video compression.

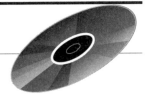

GLOSSARY

Analog

A signal which can take on any value inside a range. The most common capture format for real-world events such as video or sound. For example, full-motion video captured on film or videotape is analog, as is sound captured on a tape.

Analog to Digital Conversion

The process of converting analog video frames to digital format. Requires an analog source such as laserdisc or VCR that feeds analog video signals into the computer, and a frame grabber to isolate and convert each individual frames to digital format. Also called video capture and digitization.

Bandwidth

The throughput, or ability to move information through or from a device, system or subsystem. Usually measured in quantities of data per second. For example, single-spin CD-ROM players stream information into the computer at a maximum rate of 150 MB/second. The bandwidth of these devices is 150 kB/second.

Capture

The process of converting analog video frames to digital format. Requires an analog source such as laserdisc or VCR that feeds analog video signals into the computer, and a frame grabber to isolate and convert each individual frames to digital format. Also called analog to digital conversion and digitization.

Compression

Various algorithms, techniques and technologies applied to digital video sequences to reduce the amount of data required to recreate the original video upon decompression. Most video compression involves both interframe compression and intraframe compression techniques.

Compression Frame Rate

The number of frames per second contained in the compressed video file. Compare to display rate, which is the number of frames per second displayed during video playback.

Data Rate

Quantity of data over time required to support playback of a digital video stream. For example, uncompressed captured at a resolution of 320x240x24-bit and 30 frames per second requires a data rate of 6.9 MB/second.

Delta Frames

Frames created during interframe compression which consist solely of information describing how such frames are different from key frames. Or, frames upon which both intraframe and interframe compression may be performed. Contrast with key frames, which are defined without reference to other frames. Also called difference frames.

Digital

Information coded in binary format, or zeros and ones, usually for storage and manipulation on a digital computer.

Digital Video

Full-motion video, like that found on television and in the movies, playing in digital form on a computer. For the purposes of this title, the term specifically excludes computer-generated video sequences such as animation. Digital video should not be confused with analog video played directly on the computer through the use of special expansion cards.

Digitize or Digitization

The process of converting analog video frames to digital format. Requires an analog source such as laserdisc or VCR that feeds analog video signals into the computer and a frame grabber to isolate and convert the individual frames to digital format. Also called analog to digital conversion and digital capture.

DCI

Display Control Interface. A joint Intel/Microsoft specification designed to make video decompression and display more efficient. At its most basic level, DCI is a standard mechanism to avoid the Windows Graphics Device Interface, or GDI, which is a real drag on video peformance. Higher levels transfer decompression and display functions from the host CPU to the video card processor, where they can be performed more efficiently.

Display Rate

The number of frames per second of video displayed. A measure of system and codec peformance. Compare to compression frame rate, which is the number of frames per second actually contained in the video file.

Frame Grabber

An add-in board that captures analog signals and converts them to digital format. Also called a real-time capture card when capable of digitizing video input at 30 frames per second.

Frame Rate

A measure of video speed, usually expressed in number of frames per second or fps. Every video has two frame rates; the compression frame rate (as in, we captured and compressed at 30 fps); and the playback frame rate, which will vary by video content, compression technology, and playback platform (as in, it played at 30 fps on the '486 local bus machine, but only at 8 fps on the '386SX ISA bus machine).

Hardware Codecs

Video codecs that typically require hardware during decompression. Examples include MPEG, Motion JPEG and DVI.

Interframe Compression

Interframe compression is compression achieved by eliminating redundant information between frames. Most video sequences contain pixels that remain static over time. In most talking head videos, for example, little changes between frames except for the speaker's face and shoulders. Interframe compression techniques store the background data once, and let multiple frames access the data during decompression.

Interleave

Refers to the storage of the audio and video portions of the video stream in one combined file. This is useful for CD-ROM playback because it's more efficient to retrieve one interleaved data stream than to perform the multiple seeks required to retrieve separate audio and video streams. The Video for Windows AVI file format stands for Audio/Video Interleave.

Intraframe Compression

Intraframe compression is performed solely with reference to information within a frame. While intraframe techniques are often the most hyped, overall codec performance relates more to interframe efficiency than intraframe. Note that Indeo and Cinepak, while both vector-quantization based, perform very differently.

Key Frame

Digital video frames which are defined without reference to other frames. Or, frames upon which Intraframe Compression is performed, but not Interframe Compression. Contrast with delta frames, sequenced after a key frame, which consist solely of information describing how such frames are different from key frames. Also called reference frames because delta frames refer back to key frames during the decompression process.

Lossless

A compression scheme that reproduces files that decompress to the exact same file as the original, bit for bit. Most file compression schemes such as STAC, PKZIP and DoubleSpace are lossless because any change in the digital makeup renders most EXE and data files useless. Lossless techniques typically produce maximum compression ratios of around two to one.

Lossy

Lossy compression schemes, used primarily on video and still bitmapped images, produce files that decompress into a facsimile of the original image rather than a digitally exact replica. Lossy techniques can produce compression ratios in excess of 200:1. "Loss" begins as subtle changes in shades and increases to blockiness and artifacting at high compression ratios. A key measure of codec performance is the quality retained at high compression ratios.

Media Control Interface (MCI)

An Applications Programmer's Interface (API) for controlling multimedia devices such as laserdiscs and VCRs, as well as multimedia elements such as audio, video and animation. The specification was jointly released by

Microsoft and IBM in 1991. MCI controls are similar to BASIC commands, and the interface was primarily designed for use by programmers.

Palette

8-bit displays are limited to 256 colors. This collection of 256 colors is called the palette, since all screen elements must be painted with colors contained in the palette. The 256 color combination is not fixed—palettes can and do frequently change. But at any one point, only 256 colors can be used to describe all the objects on the screen.

Primary Target System

The target system and playback medium for which the video configuration is optimized.

Real-Time Capture

The process of capturing and digitizing video at 30 frames per second. Real-time capture products are frame grabbers that operate in real time. In addition to digitizing video, most real-time capture products also compress the incoming video in order to move the captured data through the bus to hard drive for storage.

Resolution

a) Number of pixels in a file, or displayed on a video screen. b) When applied to digital video, refers to the three-dimensional measure of the frame composition. Expressed as a series of three numbers, such such as 320x240x24, where the first number represents the number of pixels in the horizontal axis, the second number the number of pixels in the vertical axis and the third number the number of pixels describing the color of the pixel. c) When applied to a video screen, refers to the active number of pixels comprising the horizontal and vertical axis of the display.

Scale

The process of shrinking the resolution (including color depth) of analog video from its original digital resolution of approximately 512x480x24 to the target compressed digital resolution. For example, when the target compressed resolution is 320x240x8-bit, the video is digitized at 512x480x24 and then scaled to 320x240x8-bit. While not technically a compression technique, scaling has a compression-like effect by reducing the amount of data required to recreate the newly scaled video upon playback. This reduction in data is proportional to the reduction in screen resolution. For example, the original data rate of 512x480x24-bit data is approximately 22

MB/second. After scaling to 320x240x8-bit, the data rate is approximately 2.3 MB/second.

Step-Frame Capture

Non-real-time video capture. Requires a frame-accurate analog source such as laserdisc or CVD-1000 Hi-8 video deck and MCI software driver to control the process. During step-frame capture, the analog source feeds the frame when requested by the capture board. This allows the board to capture without compression, since digitization and transfer to the hard disk occur only as fast as the system can handle them.

Video Co-processor

Video co-processors are chips on graphics boards that scale and filter video to larger resolutions using proprietary techniques that enlarge without the jagged effect produced by simple pixel replication. Because these chips are designed specifically for these functions, they perform these tasks much faster than normal CPUs. Since the scaling occurs *after* the video is on the video card, it doesn't necessitate additional CPU intensive bus transfers. Example chips include VideoLogic's PowerPlay chip, available on VideoLogic and Matrox products, the Weitek Video Power and several chips from Auravision.

Video for Windows

A multimedia architecture and application suite launched by Microsoft in November of 1991. Video for Windows provides an outbound architecture that lets applications developers access audio, video and animation from many different sources through one interface. Video for Windows also provides an inbound architecture that ensures that various multimedia technologies, including codecs, work together to form a comprehensive multimedia creation and playback platform. As an application, Video for Windows primarily handles video capture and compression, and video and audio editing, and it supplies four of the five codecs discussed in this article.

ZoomDAC

Digital to analog converter (DAC) that also zooms video. All video cards have DACs that convert digital signals to analog format for sending to the monitor. ZoomDACs also scale video to higher resolutions. Because the scaling occurs at the very last point in the video display process, the additional bandwidth relating to the scaled video doesn't create any system bottlenecks. In contrast, video co-processing chips scale in video RAM, which must be processed by the DAC before display.

INDEX

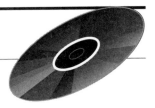